Sebastian Heidenreich

Flow Properties of Polar and Non-Polar Hard-Rod Fluids

Sebastian Heidenreich

Flow Properties of Polar and Non-Polar Hard-Rod Fluids

A study of the orientational effect on the flow behavior by the investigation of nonlinear equations for the alignment tensor and dipole vector

Südwestdeutscher Verlag für Hochschulschriften

Impressum/Imprint (nur für Deutschland/ only for Germany)
Bibliografische Information der Deutschen Nationalbibliothek: Die Deutsche Nationalbibliothek verzeichnet diese Publikation in der Deutschen Nationalbibliografie; detaillierte bibliografische Daten sind im Internet über http://dnb.d-nb.de abrufbar.
Alle in diesem Buch genannten Marken und Produktnamen unterliegen warenzeichen-, marken- oder patentrechtlichem Schutz bzw. sind Warenzeichen oder eingetragene Warenzeichen der jeweiligen Inhaber. Die Wiedergabe von Marken, Produktnamen, Gebrauchsnamen, Handelsnamen, Warenbezeichnungen u.s.w. in diesem Werk berechtigt auch ohne besondere Kennzeichnung nicht zu der Annahme, dass solche Namen im Sinne der Warenzeichen- und Markenschutzgesetzgebung als frei zu betrachten wären und daher von jedermann benutzt werden dürften.

Verlag: Südwestdeutscher Verlag für Hochschulschriften Aktiengesellschaft & Co. KG
Dudweiler Landstr. 99, 66123 Saarbrücken, Deutschland
Telefon +49 681 37 20 271-1, Telefax +49 681 37 20 271-0, Email: info@svh-verlag.de
Zugl.: Berlin, TU, Diss., 2009

Herstellung in Deutschland:
Schaltungsdienst Lange o.H.G., Berlin
Books on Demand GmbH, Norderstedt
Reha GmbH, Saarbrücken
Amazon Distribution GmbH, Leipzig
ISBN: 978-3-8381-0596-3

Imprint (only for USA, GB)
Bibliographic information published by the Deutsche Nationalbibliothek: The Deutsche Nationalbibliothek lists this publication in the Deutsche Nationalbibliografie; detailed bibliographic data are available in the Internet at http://dnb.d-nb.de.
Any brand names and product names mentioned in this book are subject to trademark, brand or patent protection and are trademarks or registered trademarks of their respective holders. The use of brand names, product names, common names, trade names, product descriptions etc. even without a particular marking in this works is in no way to be construed to mean that such names may be regarded as unrestricted in respect of trademark and brand protection legislation and could thus be used by anyone.

Publisher:
Südwestdeutscher Verlag für Hochschulschriften Aktiengesellschaft & Co. KG
Dudweiler Landstr. 99, 66123 Saarbrücken, Germany
Phone +49 681 37 20 271-1, Fax +49 681 37 20 271-0, Email: info@svh-verlag.de

Copyright © 2009 by the author and Südwestdeutscher Verlag für Hochschulschriften Aktiengesellschaft & Co. KG and licensors
All rights reserved. Saarbrücken 2009

Printed in the U.S.A.
Printed in the U.K. by (see last page)
ISBN: 978-3-8381-0596-3

Contents

I Introduction   3
    0.1 Flow Dynamics of Hard-Rod Fluids Revisited 6
    0.2 Motivation for the Present Work 7
    0.3 Outline 7

II Theoretical Foundations   11

1 Non-Polar Hard-Rod Fluids   13
    1.1 Description of the Orientation 13
       1.1.1 Orientational Distribution 13
       1.1.2 The Second-Rank Alignment Tensor 16
    1.2 Isotropic-Nematic Phase Transition 20
    1.3 Hydrodynamic Equations 22
       1.3.1 Relaxation Equation for the Alignment Tensor 22
       1.3.2 Constitutive Equation for the Pressure Tensor 23
    1.4 Flow Geometry 25
    1.5 Scaled Variables 25
       1.5.1 Relaxation Time Scaling 25
       1.5.2 Shear Rate Scaling 27
    1.6 Amended Landau-de Gennes Potential 28
       1.6.1 Theoretical Motivation 29
    1.7 Component Form of the Model Equations 32
    1.8 Further Models and Approaches 33

2 Polar Hard-Rod Fluids   37
    2.1 Orientational Distribution and its Tensorial Representation 37
    2.2 Extended Potential Function for Polar Hard-Rod Fluids 38
    2.3 Relaxation Equation and Constitutive Pressure Tensor Equation for Polar Hard-Rod Fluids 41
    2.4 Scaled Variables 44
    2.5 Component Form of the Model Equations 46
    2.6 Magnetic Fields 47

III Applications   49

3 Orientational Bulk Dynamics of Non-Polar Hard-Rod Fluids   51
       3.0.1 Extensional Flows 51
    3.1 Review of the Characteristic Solutions for the Orientational Dynamics 55
    3.2 Robustness of Periodic and Chaotic Solutions 57

CONTENTS

		3.2.1	Modeling of Shear Rate Perturbations	57
		3.2.2	Isotropic Phase, Flow Alignment and Periodic Solutions	60
		3.2.3	Chaotic Solutions	64

4 Spatially Inhomogeneous Dynamics of Non-Polar Hard-Rod Fluids   **69**
- 4.1 Equilibrium States 69
- 4.2 Apparent Slip of the Isotropic State Subjected to a Flow 70
  - 4.2.1 Boundary Conditions 72
  - 4.2.2 Isotropic Phase and Small Shear Rates 73
  - 4.2.3 One-Dimensional Spatial Dependence 75
  - 4.2.4 Plane Couette Flow 76
  - 4.2.5 Plane Poiseuille Flow 80
  - 4.2.6 Flow Down an Inclined Plane 84
  - 4.2.7 Alignment 87
  - 4.2.8 Cylindrical Couette Flow Geometry 88
- 4.3 Orientational Dynamics and Flow Properties of Nematic State 92
  - 4.3.1 Imposed Shear 92
  - 4.3.2 Hydrodynamics: Oscillating Jet-Layers 101
  - 4.3.3 Oblate Defects and Jet-Generation Mechanism 101
  - 4.3.4 Multiple Jets and Scaling Behavior 105

5 Spatially Inhomogeneous Dynamics of Polar Hard-Rod Fluids   **111**
- 5.1 Shear-Induced Dynamic Polarization and Mesoscopic Structure 111

6 Summary, Conclusions and Outlook   **121**
- 6.1 Summary and Conclusions 121
- 6.2 Outlook for Further Investigations 124

7 Appendix   **125**
- 7.1 Numerics 125
- 7.2 The Probability Distribution Function for Polar Hard-Rod Fluids 127

Part I

Introduction

The term flowing crystal and liquid crystal was introduced by Otto Lehmann over 120 years ago [1–5]. The first thermotropic liquid crystal was found by Reinitzer [6]. He sent the substance to Lehmann, who noticed the birefringence in the liquid state. Since before birefringence had only been observed in crystals the somewhat contradictionary expression liquid crystal was introduced and is still in use today. In the meantime many new phases between the solid and ordinary isotropic liquid state are known. They are referred as to mesophases.

Liquid crystalline mesophases posses ordinary properties of liquids but, at the same time, show anisotropy in their mechanical and electromagnetic properties.

Molecules showing liquid crystalline phases are typically shaped as rods or disks. One distinguishes between thermo- and lyotropic liquid crystals [7]. Thermotropic liquid crystals are mesomorphic in a certain temperature range and lyotropic in a certain concentration range, respectively. The most common liquid crystalline mesophases are nematic and smectic [7].

In the nematic phase there is a tendency of the molecules to orient parallel to each other such that one direction is preferred. On the other hand the smectic phase is characterized by an additional layering structure. The simplest liquid crystalline phase is the uniaxial nematic phase, its orientational distribution is uniaxial in equilibrium. The appropriate order parameter for the description of uniaxial as well as biaxial orientations is the second rank symmetric traceless tensor **a** referred to as alignment tensor (first non vanishing moment of the orientational distribution function). It can be detected directly by birefringence experiments.

The preferred orientation of nematic liquid crystals is commonly described by a unit vector referred to as *director*. However, in non-equilibrium (e.g. shear flow) the orientation is no longer uniaxial and becomes biaxial. Therefore the full alignment tensor description is necessary even in the case where the equilibrium state is isotropic.

The generic model used here ignores the molecular details and is applicable to fluids that in principle consists of hard rods or hard disks. Representative examples are liquid crystals (low molecular weight liquid crystals as used in liquid-crystal displays), nano-composites, liquid crystal polymers, worm-like micelles, tobacco mosaic virus suspensions and inorganic nano-crystals [8–15]. In each of these materials, orientational degrees of freedom and the possibility to form different mesoscopic phases (isotropic and nematic) leads to surprising and fascinating flow phenomena [16–24]. In the last decade the experimental technics of precise designing and synthesizing nano-rods was successfully developed [25] and properties like the strength of the electric or magnetic dipole moment, the aspect ratio and the shape of nano-rods are controllable [26–28]. For applications, flow controlling by molecular features is highly desirable and theoretical investigations of the orientational behavior are beneficial.

Many hard-rod fluids (liquid crystals) show a specific symmetry of its orientational distribution function. The orientational distribution function is invariant under the local rotational transformation of every molecule by multiple of π. The symmetry manifest in the Cartesian tensorial expansion of the probability distribution function. In the expansion the first non-vanishing moment is the second rank tensor (alignment tensor). However, in general there are fluids showing polarized phases in addition to the nematic phase (ferronematics). For this class of complex fluids an additional order parameter (dipole vector) is necessary. In this work both kinds of fluids are investigated. Non polar hard-rod fluids possessing the head-tail symmetry (alignment tensor order parameter) whereas polar hard-rod fluids are not (with additional dipole vector order parameter).

0.1 Flow Dynamics of Hard-Rod Fluids Revisited

Non-polar hard-rod fluids (nematic liquid crystals) subjected to a shear flow respond with a time-dependent orientational behavior or stationary flow alignment. The time-dependent phenomena can be rather complex. Different types of spatially homogeneous periodic behavior referred to as (ordinary) tumbling, wagging, have been identified in experiments and in theoretical descriptions [29–32]. In particular, the long-time transient kayaking motion have been identified in experiments [32–35] and confirmed through theoretical descriptions [36–44].

A relatively simple model based on a nonlinear equation for the second rank alignment tensor (introduced by Hess [45–47]) could confirm the oscillatory and flow alignment flow response. In addition the model reveals a more complex and even chaotic behavior for certain model parameters and specific values of the applied shear rate [43, 44]. Chaotic behavior was also found from a solution of a Fokker-Planck equation for the orientational distribution function involving 65 components rather than the 5 independent components of the second rank alignment tensor [48]. Chaotic solutions arise through a period-doubling bifurcation route, which Berry [34] associated with the rapid development of turbidity in experiments. It is in this flow regime where the homogeneity assumption of the orientational distribution and the presumption of steady, linear shear become especially suspect, is a strong motivation to undertake spatio-temporal numerical studies. Further theoretical studies on the periodic and chaotic orientational and rheological behavior are presented in [16, 24, 42, 49].

The homogeneous flow response of polar hard-rod fluids cover in general the steady, oscillating and chaotic solutions observed for non-polar fluids. In addition there is a wide range of new characteristic solutions, e.g. for, transient and in-plane chaotic states. In addition, the orientational dynamics strongly depends on the cases where the dipole vector is parallel or perpendicular to the molecular axes [50–52].

For spatially inhomogeneous systems with physical boundary conditions on parallel, oppositely moving plates, models continue to yield transitions between regular and complex spatio-temporal behavior, including persistence of chaotic dynamics [24, 53–56], and shear banding [17, 18]. In some studies, the flow is imposed as simple linear shear and orientational gradients are allowed, while other, more resolved simulations perform a self-consistent computation of the flow. In both flow-imposed and flow-coupled simulations, a rich phase diagram of heterogeneous space-time attractors is predicted [41, 57–59], where once again most attention has been given to the orientational distribution. This is true not just because the full hydrodynamic coupling to the Navier-Stokes momentum equation is so numerically challenging, but also because of the lack of experimental resolution of flow inside the shear cell to benchmark model predictions.

A particularly interesting flow feedback phenomenon with a compelling experimental evidence for the formation of steady roll cells, two-dimensional secondary flows in the shear-gradient and vorticity fields, at very low shear rates. These structures were reported experimentally by Larson and Mead [60, 61], and successfully modeled by Feng, Tao & Leal [19] with a liquid crystal director theory, and more recently by Klein et al. [62].

Similar phenomenological models for complex fluids are derived and investigated within principles of continuum mechanics and non-equilibrium thermodynamics [22, 63–68]. For models using the Poisson-bracket approach it is refered to [69].

0.2 Motivation for the Present Work

Hard-rod fluids are a general and simple model for a wide range of anisotropic, non-Newtonian fluids that consist of small-to-large molecules with properties similar to rigid rods or platelets. In each of these model systems, orientational degrees of freedom and the possibility to form different mesoscopic phases (isotropic and nematic) leads to surprising orientational behavior and flow feedback in shear-dominated rotational flows. The large literature on shear banding [17, 18] in sheared worm-like micelles [70] is an example of the remarkable non-Newtonian flow feedback that is possible in such systems.

In addition, for microfluidic length scales physical conditions at the confining walls impose microstructure and strongly affect the flow properties. Especially, the apparent slip caused by the molecular interaction with the solid surface was the reason for many theoretical and experimental studies, see for example [71–76].

For application in microfluidic devices boundary conditions are important for understanding the flow properties. The motivation of this work was to investigate the influence of boundaries on the orientational dynamics and on the flow properties. It was expected that the competition between boundary induced mesoscopic structures and hydrodynamic orientational behavior yields new fascinating non-Newtonian flow feedback phenomena relevant for microfluidic applications.

As a starting point, this work used a relatively simple model for non-polar had-rod fluids [45–47]. The aim was to extend the bulk dynamics results [43, 44, 77–79] to the spatially inhomogeneous systems with non-Newtonian flow feedback.

The second aim was to extend and investigate the theoretical model for polar hard-rod fluids. It was expected that rather interesting and surprising orientational-flow effects and application for microfluidic devices arise due to additional non-vanishing average dipole moment.

0.3 Outline

Chapter 2

The second chapter provides the theoretical background for this work. In the first part the alignment tensor is introduced and the relaxation equations as well as the constitutive equations for the pressure tensor are given. For numerical studies scaled variables are introduced. Furthermore, a theoretical motivation for an amended nematic potential compared to the frequently used Landau-de Gennes potential is presented.

Here the Landau-de Gennes free energy which includes terms up to 4th order in the alignment tensor and which does not impose a upper bound on the magnitude of the alignment tensor was amended by a version which includes arbitrary high orders and does impose a realistic bound. This point is of importance for numerical solutions, in particular in spatially inhomogeneous situations where run-away solutions might lead to unphysically large values of the alignment.

In the second part of the chapter the spatially inhomogeneous relaxation equations and the constitutive pressure equation for polar hard-rod fluids are derived and scaled variables are introduced. Dynamic polarization leads to the occurrence of magnetic fields. Based on Maxwells equations the equations for polarization-induced magnetic fields are derived and presented.

Chapter 3

First, the effect of the amended potential on the order parameter in comparison to the order parameter behavior involving the Landau-de Gennes potential is studied. For the simple shear flow, there are (small) quantitative changes of the parameter ranges where the various types of the orientational behavior is found. This changes strongly for extensional flows. For extensional flow the order parameter increase with no bounds if the Landau-de Gennes potential is used. On the other hand if the amended potential is used the order parameter is restricted and agrees with experimental observations.

In the second part of the chapter the characteristic solutions of the orientational dynamics in the bulk system subjected to a steady shear are revisited. Based on a numerical analysis the robustness of the homogeneous solutions against perturbations in the shear rate are investigated. It is demonstrated that periodic and chaotic solutions can be surprisingly robust against such distortions.

Chapter 4

The boundary conditions are formulated for the second rank alignment tensor describing the orientation of non-polar hard-rod fluids and for the velocity slip. The guiding principle, in the spirit of irreversible thermodynamics, is the same as that originally suggested for gases [80] viz.: i) the entropy production at an interface is inferred from the entropy flux in the bulk fluid, ii) the boundary conditions are set up such that the interfacial entropy production is positive definite. The extension to molecular gas and to molecular liquids was presented in [81, 82]. For isothermal flow of non-polar hard-rod fluids in the isotropic phase, it is demonstrated that the coupling between the alignment tensor and the friction pressure tensor leads to an apparent velocity slip even when the velocity obeys a stick boundary condition. The velocity and alignment profiles, as well as the effective viscosities are calculated for plane and cylindrical Couette and plane Poiseuille flow, as well as the flow down an inclined plane. The dependence of these quantities and of the apparent slip velocity on a microscopic length parameter and on the ratio between the first and second Newtonian viscosities are discussed. In experiments slip lengths and the viscosities of thin films of Newtonian liquids were measured and studied by Jacobs et al. [83]. Furthermore, a recent thermodynamic formulation of boundary conditions building upon the pioneering work of Waldmann [80] was derived in [84–86].

The second part of the chapter deals with the flow feedback behavior in the strongly nonlinear regime, where both anisotropy and focusing-defocusing of the orientational distribution are important. In the results reported here, attractors that are unsteady in both flow and orientation, heterogeneous in one space dimension, and yet the orientational distribution is approximately in-plane. In this window of bulk shear rates, the nonlinear flow feedback phenomenon consists of oscillating or pulsating jet-like layers. Some scaling properties of the localized jet layers, such as where they reside in the shear gap, are given with respect to Deborah and Ericksen numbers. This strongly nonlinear behavior is impossible with a pure director theory such as Leslie-Ericksen-Frank theory, which does not allow order parameter degrees of freedom nor biaxiality. A similar effect was found for a planar "two dimensional liquid" model studied and reported by Kupferman *et al* [87]. In principle, the non-linear flow feedback effect is also present in other models for non-polar hard-rod fluids as investigated and reported in [88].

Chapter 5

In the last chapter of this work the flow behavior of a class of complex fluids composed of fluids with permanent dipole moments (polar hard-rods) is considered. It is shown that spontaneous polarization can occur in sheared polar hard-rod fluids with mesoscopic spatial structure. It is focused on systems where the structure is induced by the combination of shear flow and confining walls, such as in micro-channels. The investigations are based on the numerical solutions of the spatially one-dimensional hydrodynamic model including feedback, the full alignment tensor as well as the dipole vector. The study generalizes earlier approaches for homogeneous (bulk) systems of polar hard-rods [50, 51], where a non-vanishing average dipole moment only appears if the equilibrium state is ferroelectric. On the contrary, the structured systems develop spontaneous, time-dependent polarization for a wide range of parameters and boundary conditions. For time-dependent polarization magnetic fields result. The parameter dependence and possible applications of the occurring magnetic fields are discussed. Finally, this work concludes with chapter 6.

Part II

Theoretical Foundations

1
Non-Polar Hard-Rod Fluids

1.1 Description of the Orientation

1.1.1 Orientational Distribution

The orientation of a hard rod or of the backbone of a stiff molecule is specified by the unit vector $\boldsymbol{u}$. The statistical state of the fluid is characterized by the orientational distribution function $\tilde{\rho}^{\text{or}}(u_\nu, r_\mu, t)$. The probability to find a molecule in the interval $[r'_\mu, r'_\mu + dr_\mu]$ with the orientation $[u_\nu, u_\nu + du_\nu]$ at the time t is given by $\tilde{\rho}^{\text{or}}(u_\nu, r_\mu, t) d^3r d^2u$.

For a mesoscopic description the system is coarse-grained by averaging over the interval $[r'_\mu, r'_\mu + dr_\mu]$. Here it is assumed that in the interval the system is spatially homogeneous, such that for every quantity $A(r'_\mu + dr_\mu) = A(r'_\mu)$. The vector $\boldsymbol{r}'$ denotes the coarse-grained position vector. When the position $\boldsymbol{r}'$ and the time t is fixed the orientational distribution function ρ^{or} is defined on the unit sphere S^2 and satisfies the normalization condition

$$\int_{S^2} \rho^{\text{or}}(\boldsymbol{u}, \boldsymbol{r}', t) d^2u = 1. \tag{1.1}$$

The ensemble average of A at $(\boldsymbol{r}', t)$ is given by

$$\langle A(\boldsymbol{u}) \rangle (\boldsymbol{r}', t) = \int_{S^2} \rho^{\text{or}}(\boldsymbol{u}, \boldsymbol{r}', t) A(\boldsymbol{u}) d^2u. \tag{1.2}$$

The symbol d^2u stands for the solid angle element on the unit sphere. In polar angles (φ, θ) the components u_μ, with $\mu = 1, 2, 3$ are given by

$$u_1 = \sin\theta \cos\varphi \ , \ u_2 = \sin\theta \cos\varphi \ , \ u_3 = \cos\theta \tag{1.3}$$

and therefore $d^2u = \sin\theta d\theta d\varphi$ as determined by the determinant of the corresponding metric. In the following considerations are made for fixed $(\boldsymbol{r}', t)$.

Generally, functions on the unit sphere can be expanded with respect to spherical harmonics $Y_\ell^{(m)}(\boldsymbol{u}) = Y_\ell^{(m)}(\theta, \varphi)$. Spherical harmonics forms a complete set of orthogonal functions. The expansion of the orientational distribution function is given by [89–92]

$$\rho^{\text{or}}(\boldsymbol{u}) = \sum_{\ell=0}^{\infty} \sum_{m=-\ell}^{\ell} \langle Y_\ell^{(m)}(\boldsymbol{u}) \rangle Y_\ell^{(m)*}(\boldsymbol{u}). \tag{1.4}$$

CHAPTER 1. NON-POLAR HARD-ROD FLUIDS

The asterisks denotes the complex conjugation of the complex-valued function $Y_\ell^{(m)}$. Otherwise spherical harmonics Y_ℓ can be related to the irreducible part of Cartesian symmetric tensors of rank ℓ [90–92]. In the special basis

$$\mathbf{e}^{(0)} = \mathbf{e}_z, \quad \mathbf{e}^{(\pm 1)} = \mp \frac{1}{2}(\mathbf{e}_x \mp i\mathbf{e}_y) \tag{1.5}$$

spherical harmonics can be expressed, viz

$$\left(\overline{\boldsymbol{u}^\ell}\right)_m = \sqrt{\frac{4\pi \ell!}{(2\ell+1)!!}} Y_\ell^{(m)}(\boldsymbol{u}). \tag{1.6}$$

Here u^ℓ denotes the ℓ-fold dyadic product of the unit vector $\boldsymbol{u}$ and the symbol $\overline{\mathbf{x}}$ indicates the symmetric traceless part of a tensor $\mathbf{x}$ (irreducible part), i.e. with Cartesian components denoted by Greek subscripts, one has $\overline{x_{\mu\nu}} = (1/2)(x_{\mu\nu} + x_{\nu\mu}) - (1/3)x_{\lambda\lambda}\delta_{\mu\nu}$. With the abbreviations

$$\xi_\ell = \sqrt{\frac{(2\ell+1)!!}{\ell!}}, \quad a_{(\ell)} = \langle \overline{\boldsymbol{u}^\ell} \rangle, \tag{1.7}$$

Eq. (1.4) and Eq. (1.6) the expansion of the orientational distribution function with respect to Cartesian tensors yields

$$\rho^{\mathrm{or}}(\boldsymbol{u}) = \frac{1}{4\pi}\left[1 + \sum_{\ell=1}^{\infty} \xi_\ell a_{(\ell)} \otimes^{(\ell)} \overline{\boldsymbol{u}^{(\ell)}}\right]. \tag{1.8}$$

Here the symbol $\otimes^{(\ell)}$ denotes the ℓ th contraction of the tensors $a_{(\ell)}$ with the irreducible part of $\boldsymbol{u}^{(\ell)}$. Many liquid crystals and nano-rod dispersions show a specific symmetry behavior, viz the orientational distribution function is independent on the transformation $\boldsymbol{u} \to -\boldsymbol{u}$ referred to as "head-tail symmetry". Note, that the single molecule can exhibit a permanent dipole moment. As a consequence of the symmetry the orientational distribution function $\rho^{\mathrm{or}}(\boldsymbol{u},\boldsymbol{r}',t)$ is defined on the projective plane PS^2 for fixed $(\boldsymbol{r}',t)$, i.e. a sphere where the antipodes are identified. The head-tail symmetry $\rho^{\mathrm{or}}(\boldsymbol{u},\boldsymbol{r}',t) \stackrel{!}{=} \rho^{\mathrm{or}}(-\boldsymbol{u},\boldsymbol{r}',t)$ is responsible for the occurrence of terms only with even ℓ in the expansion (1.8)

$$\rho^{\mathrm{or}}(\boldsymbol{u}) = \frac{1}{4\pi}\left[1 + \sum_{\ell=1}^{\infty} \xi_{2\ell} a_{(2\ell)} \otimes^{(2\ell)} \overline{\boldsymbol{u}^{(2\ell)}}\right]. \tag{1.9}$$

In principle, the Eq. (1.9) specify an expansion of ρ^{or} with respect to its moments. The first nontrivial moment

$$\mathbf{a}_{(2)} = \sqrt{\frac{15}{2}} \langle \overline{\boldsymbol{u}\boldsymbol{u}} \rangle \tag{1.10}$$

is the second rank symmetric traceless tensor referred to as (second rank) *alignment tensor*. The equivalent expansion can be made for every $(\boldsymbol{r}',t)$ yielding to an spatially and time-dependent mesoscopic order parameter $\mathbf{a}$. In Fig. 1.1 the hole coarse graining procedure is illustrated.

1.1. DESCRIPTION OF THE ORIENTATION

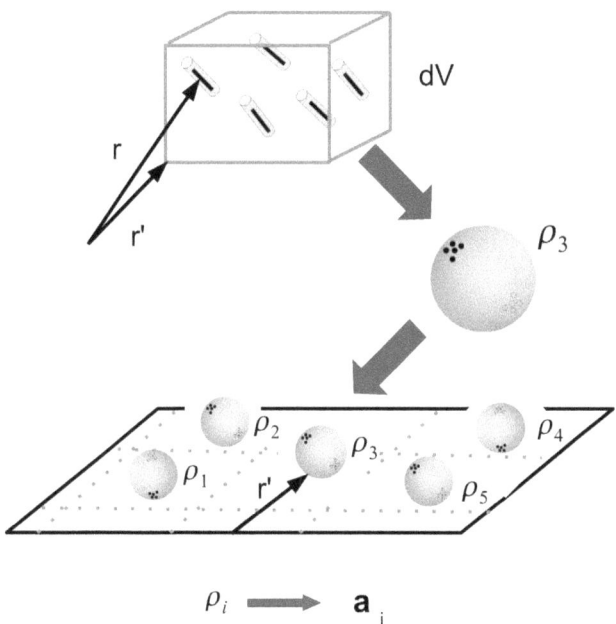

Figure 1.1: A system characterized by the orientational distribution function $\tilde{\rho}$ is considered. For the determination of a mesoscopic description the volume dV is related to a specific length scale where fluctuations are less important. Every volume dV_i is considered to be specially homogeneous and the molecules inside determine a homogeneous probability distribution function ρ_i. The orientational distribution function ρ_i is characterized by its first nontrivial moment, i.e. the alignment tensor $\mathbf{a}_i$. This means the alignment tensor $\mathbf{a}(r_\mu)$ represents the local orientation of the corresponding mesoscopic volume.

CHAPTER 1. NON-POLAR HARD-ROD FLUIDS

1.1.2 The Second-Rank Alignment Tensor

Liquid crystalline phases of rod dispersions are characterized by order parameters that measure the anisotropy of the fluid. In an isotropic system the orientation is random and the distribution function independent of $\boldsymbol{u}$, i.e.

$$\rho_{\text{iso}}^{\text{or}} = \frac{1}{4\pi}. \tag{1.11}$$

The averages of $\boldsymbol{u}$ and $\boldsymbol{uu}$ with $\rho_{\text{iso}}^{\text{or}}$ are

$$\langle \boldsymbol{u} \rangle_{\text{iso}} = 0, \quad \langle \boldsymbol{uu} \rangle_{\text{iso}} = \frac{1}{3}\delta. \tag{1.12}$$

Due to the head-tail symmetry the average of **u** is zero even for anisotropic distribution functions such that it is not an appropriate order parameter. For the description of anisotropic properties of the fluid the derivation of $\boldsymbol{uu}$ from isotropy is used. The derivation is given by $\boldsymbol{uu} - \langle \boldsymbol{uu} \rangle_{\text{iso}} = \boldsymbol{uu} - \frac{1}{3}\delta = \overline{\boldsymbol{uu}}$ and in the average it is proportional to the second rank alignment tensor (1.10)

$$\mathbf{a}(\mathbf{x},t) = \mathbf{a}(\mathbf{r},t)_{(2)} = \sqrt{\frac{15}{2}} \langle \overline{\boldsymbol{uu}} \rangle (\mathbf{r},t). \tag{1.13}$$

Physical quantities depending on the orientation $\boldsymbol{u}$ can be described in the mesoscopic description with the alignment tensor, i. e. $A(\boldsymbol{u}, \boldsymbol{r}, t) \to A(\mathbf{a}, \boldsymbol{r}', t)$. Frequently, the alignment tensor is referred to as **Q**-tensor, sometimes **S**-tensor. The factor $\sqrt{15/2}$ is convention.

In general, a second rank symmetric traceless tensor in N dimensions has $1/2N(N+1)$ independent components. For $N=3$ five components are independent. Three components can be used to fix the reference frame and the other two to describe the average orientational order. In the principal frame the tensor is represented in its diagonal form

$$\mathbf{a} = \mu_1 \mathbf{ll} + \mu_2 \mathbf{mm} + \mu_3 \mathbf{nn}, \tag{1.14}$$

where **m**, **n**, **l** are parallel to the principal axes. The coefficients μ_i are the corresponding principal values that satisfy the traceless condition $\sum_{i=1}^{3} \mu_i = 0$. In the case where the components of $\boldsymbol{u}$ are expressed in polar angles (φ, θ) cf. (1.3) the principal values are given by

$$\mu_1 = \sqrt{\tfrac{15}{2}} \left(X - \tfrac{1}{3} \right) \tag{1.15}$$

$$\mu_2 = \sqrt{\tfrac{15}{2}} \left(Y - \tfrac{1}{3} \right) \tag{1.16}$$

$$\mu_3 = \sqrt{\tfrac{15}{2}} \left(Z - \tfrac{1}{3} \right), \tag{1.17}$$

with the abbreviations

$$X = \langle (\boldsymbol{\ell} \cdot \boldsymbol{u})^2 \rangle = \langle \sin^2 \theta \cos^2 \phi \rangle \tag{1.18}$$

$$Y = \langle (\boldsymbol{m} \cdot \boldsymbol{u})^2 \rangle = \langle \sin^2 \theta \sin^2 \phi \rangle \tag{1.19}$$

$$Z = \langle \cos^2 \theta \rangle. \tag{1.20}$$

The principal values are used to characterize different states of order. If the principal values are sorted by size ($\mu_1 \leq \mu_2 \leq \mu_3$) one distinguishes between

1.1. DESCRIPTION OF THE ORIENTATION

- isotropic order: $\mu_i = 0$
- uniaxial order: $\mu_1 = \mu_2 < \mu_3$
- biaxial order: $\mu_1 < \mu_2 < \mu_3$
- planar biaxial order (frequently, referred to as plate-like defect): $\mu_1 = -\mu_3$, $\mu_2 = 0$.

As mentioned the three principal values are not independent. For two independent components (p, q) the following ansatz is made

$$\mathbf{a} = \sqrt{\frac{3}{2}} q \overline{\mathbf{nn}} + \frac{1}{\sqrt{2}} p (\mathbf{ll} - \mathbf{mm}). \tag{1.21}$$

The coefficients are chosen such that $\mathbf{a} : \mathbf{a} = p^2 + q^2$. The principal values are related to the parameters (p, q) via

$$\mu_1 = \frac{1}{\sqrt{2}} p - \frac{1}{\sqrt{6}} q, \quad \mu_2 = -\frac{1}{\sqrt{2}} p - \frac{1}{\sqrt{6}} q, \quad \mu_3 = \frac{2}{\sqrt{6}} q. \tag{1.22}$$

The parameter q is related to uniaxial alignment and the parameter p to biaxial alignment as shown in the following, respectively. The uniaxial part of the alignment tensor is projected out by contraction with $\mathbf{nn}$

$$\mathbf{nn} : \mathbf{a} = \sqrt{\frac{2}{3}} q. \tag{1.23}$$

Otherwise $\mathbf{nn} : \overline{\mathbf{uu}} = \frac{2}{3} P_2(\mathbf{n} \cdot \mathbf{u})$ such that

$$q = \sqrt{5} S_2, \tag{1.24}$$

where $S_2 = \langle P_2(\mathbf{n} \cdot \mathbf{u}) \rangle = \langle P_2(\cos\theta) \rangle$ is referred to as the *Maier-Saupe* order parameter [93–95]. The uniaxial orientational distribution function ($q \neq 0$ and $p = 0$) yields

$$\rho^{\text{or}} = \frac{1}{4\pi} \left(1 + 5 S_2 P_2(\cos\theta) \right) = \frac{1}{4\pi} \left(1 + 5 S_2 P_2(\mathbf{n} \cdot \mathbf{u}) \right), \tag{1.25}$$

where $P_2(x) = \frac{1}{2}(3x^2 - 1)$ denotes the second Legendre polynomial. The angle θ characterizes the orientation of a molecular axes $\mathbf{u}$ compared to the preferred direction indicated by the *director* $\mathbf{n}$. The uniaxial orientational distribution function is rotational invariant around the director $\mathbf{n}$.

In the case of biaxial orientation the distribution function is more complex. The contraction of the alignment tensor $\mathbf{a}$ by the tensor $\mathbf{ll} - \mathbf{mm}$ yields the biaxial order parameter and is given by

$$b = \frac{\sqrt{15}}{2} Q_2, \tag{1.26}$$

with $Q_2 = \langle \sin^2\theta \cos 2\phi \rangle$. The biaxial orientational distribution ($q \neq 0$ and $p \neq 0$) function reads

$$\rho^{\text{or}} = \frac{1}{4\pi} \left(1 + 5 S_2 P_2(\mathbf{n} \cdot \mathbf{u}) + \frac{15}{4} Q_2 \left((\mathbf{l} - \mathbf{m}) \cdot \mathbf{u} \right) \right). \tag{1.27}$$

CHAPTER 1. NON-POLAR HARD-ROD FLUIDS

Instead of one preferred direction ρ^{or} depends on two "directors" $\mathbf{n}$ and $\boldsymbol{\ell} - \mathbf{m}$.

The specific values of the order parameters (p, q) are determined by the orientational distribution function and bounded due to the normalization condition. In particular the relation $0 \leq X, Y, Z \leq 1$ bounds the values of (p, q) to

$$-\frac{\sqrt{5}}{2} \leq q \leq \sqrt{5}, \quad -\frac{\sqrt{15}}{2} \leq p \leq \frac{\sqrt{15}}{2}. \tag{1.28}$$

For uniaxial alignment the bounds are in agreement with the fact, that the Maier-Saupe order parameter is restricted to $-\frac{1}{2} \leq S_2 \leq 1$.

The scalar order parameters (q, p) are not unique since the principal axes can be interchanged cyclically. A suitable measure for the biaxiality is given by the biaxiality parameter [96]

$$b^2 = 1 - \frac{I_3^2}{I_2^3}, \tag{1.29}$$

where I_2 and I_3 denotes the second and third scalar rotational invariants of $\mathbf{a}$, respectively. The second scalar invariant is the square of the norm and the third scalar invariant the determinant of $\mathbf{a}$ (see [97]), i. e.

$$I_2 = a_{\mu\nu}a_{\mu\nu} = p^2 + q^2, \quad I_3 = \sqrt{6}a_{\mu\nu}a_{\nu\lambda}a_{\lambda\mu} = p^3 - 3pq \tag{1.30}$$

The cases $b = 0$ and $b = 1$ correspond to uniaxial and planar biaxial alignment, respectively. In Fig. 1.2 the biaxiality in the $p - q$ plane is displayed. There are several regions where the biaxiality parameter $b = 0$ indicating uniaxial alignment. At the horizontal axes $p = 0$ the distribution is uniaxial in the direction $\mathbf{n}$. For the other regions where $b = 0$ the distribution is uniaxial and preferentially shows in $\mathbf{l}$ or $\mathbf{m}$ direction. The white regions ($b = 1$) are related to planar biaxiality.

Tensor Basis

The symmetric traceless alignment tensor can be expressed in a five dimensional standard [96] ortho-normalized tensor basis

$$\mathbf{a} = \sum_{k=0}^{4} a_k \mathbf{T}^k, \tag{1.31}$$

where $\mathbf{T}^i$ with $i = 0, .., 4$ are the basis tensors by which $\mathbf{a}$ is uniquely expressed:

$$\mathbf{T}^0 \equiv \sqrt{3/2}\,\overline{\mathbf{e}^z\mathbf{e}^z}, \quad \mathbf{T}^1 \equiv \sqrt{1/2}\,(\mathbf{e}^x\mathbf{e}^x - \mathbf{e}^y\mathbf{e}^y), \quad \mathbf{T}^2 \equiv \sqrt{2}\,\overline{\mathbf{e}^x\mathbf{e}^y}, \\ \mathbf{T}^3 \equiv \sqrt{2}\,\overline{\mathbf{e}^x\mathbf{e}^z}, \quad \mathbf{T}^4 \equiv \sqrt{2}\,\overline{\mathbf{e}^y\mathbf{e}^z}. \tag{1.32}$$

The orthogonality relation and the expression for the coefficients a_k are given by

$$\mathbf{T}^i : \mathbf{T}^k = \delta_{ik} \quad \text{and} \quad a_i = \mathbf{a} : \mathbf{T}^i. \tag{1.33}$$

1.1. DESCRIPTION OF THE ORIENTATION

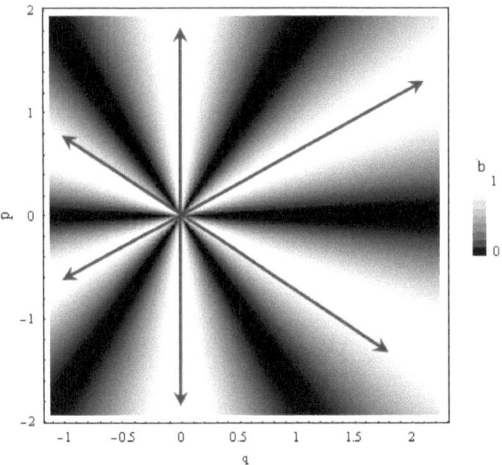

Figure 1.2: The biaxial parameter is given in the $p - q$-plane. The (blue) arrows indicates uniaxial alignment ($b = 0$) regarding to the principal axes $\mathbf{n}$, $\mathbf{m}$ and $\boldsymbol{\ell}$, respectively.

CHAPTER 1. NON-POLAR HARD-ROD FLUIDS

Visualization of Second Rank Tensors

The visualization of the alignment tensor **a** is very useful for the interpretation of the orientational behavior in the flow. For the visualization of **a** different geometric approaches are possible. On one hand the tensor is visualized by small bricks and on the other hand by ellipsoids. In both cases the eigenvalues and the eigenvectors of the alignment tensor determine the graph.

In the brick picture [44, 68] the orientation of the brick is given by the trihedral orientation formed by the three eigenvectors of the alignment tensor. The shape of the brick is characterized by the corresponding eigenvalues. Every edge of the brick is related to one eigenvector. The length of the edge is equal to the eigenvalue of the related eigenvector. To avoid vanishing bricks (as in the isotropic case) one add $1/3$ to the eigenvalues. The length of the edges are

$$\ell_i = \sqrt{\frac{2}{15}\mu_i + \frac{1}{3}}. \qquad (1.34)$$

For information about the strength of order the bricks are colored. White color denotes minimum values of $|a| = |\sqrt{\mathbf{a} : \mathbf{a}}|$ and black maximal values, respectively.

In the ellipsoid description [98] the eigenvalues d_i are scaled to obey $d_1 + d_2 + d_3 = 1$ and ordered according to $0 < d_3 < d_2 < d_1 < 1$. The orientation of the ellipsoids is given by the trihedral of the eigenvectors and the shape by the quadratic form $Q(\mathbf{x}) = 1$, where $Q = \mathbf{x}^T A \mathbf{x}$ and $A = diag(d_1, d_2, d_3)$. The quadratic form represents a surface. Here it is an ellipsoid with the axes length (d_1, d_2, d_3).

1.2 Isotropic-Nematic Phase Transition

Liquid crystals and rod dispersions are characterized by microscopic orientational order. Dependent on the temperature (thermotropic) or the concentration (lyotropic) it shows different phases. For high temperature (low density) molecules are disordered: isotropic phase. If the temperature decreases (or density increases) the aligned state is energetically more favored at a critical value, i.e. the orientation prefers one direction whereas the positions are disordered. This state is referred to as the nematic phase. The preferred direction defines the director.

Beside the isotropic and nematic phase there are many more mesophases between the solid and liquid phase. In the cholesteric phase a director can be defined in planes, say the xy-plane. In the z- direction the director rotates and draws a helix with the cusp of the director. Furthermore, different smectic phases can be identified. In smectic phases the molecules forms layers. In the layers the director indicates a preferred direction and the molecules are positionally disordered (fluid behavior, see Fig. 1.3). Here the focus is on the isotropic- to nematic phase transition. The description of further phase transitions is e.g. found in [99]. In the spirit of Landau's phenomenologically description of second order phase transitions, de Gennes developed a theory that describes the first order isotropic-to nematic phase transition. A reasonable order parameter is the symmetric traceless tensor **a**. That vanishes in the isotropic phase and is unequal to zero in the nematic phase. The second rank alignment tensor is closely related to the birefringence which distinguishes the nematic from

1.2. ISOTROPIC-NEMATIC PHASE TRANSITION

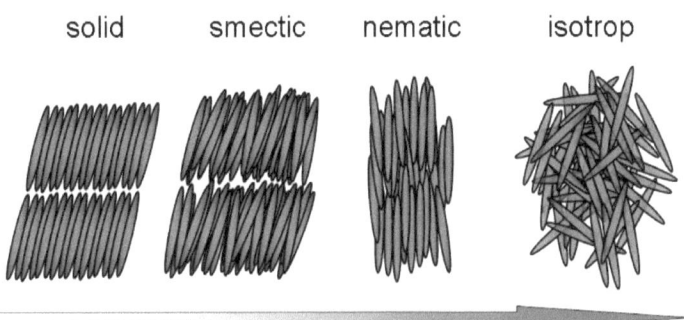

Figure 1.3: Different phases between the solid and liquid phase can occur in rod suspensions. The solid is characterized by orientational and positional perfect order. For high temperature the isotropic fluid phase without positional and orientational order is observed. In between many mesophases can be exist, eg. the smectic C (layering) and the nematic phase (positional disorder, orientational order) is shown.

the isotropic phase. In general, the free energy F is a scalar and the expansion of F in powers of the order parameter $\boldsymbol{a}$ contains terms that are invariant against rotations of the reference frame. The general form of the free energy functional is constructed from the scalar rotational invariants of the order parameter [97], i.e.

$$F = \sum_m \sum_{n_1...n_m} c_{n_1...n_m} \prod_{i=1...m} I_{n_i}, \quad n_\alpha = 1, 2..., \quad m = 1, 2... \quad (1.35)$$

with the n_i-th invariant

$$I_{n_i} = Tr(\boldsymbol{a}^{n_i}). \quad (1.36)$$

In the case $N = 3$ it can be shown that all invariants higher than I_3 can be expressed as polynoms of I_1, I_2, I_3 [97]. The traceless condition gives $I_1 = 0$ and the free energy consists only of combinations of the second $I_2 = a_{\mu\nu}a_{\mu\nu}$ and third $I_3 = a_{\mu\lambda}a_{\lambda\nu}a_{\nu\mu}$ rotational scalar invariant. The expansion up to the 4th order leads to the Landau-de Gennes potential (which proportional to the free energy) [7]

$$\Phi^{LDG} = (1/2)A(T)a_{\mu\nu}a_{\mu\nu} - (1/3)\sqrt{6}\, B\, a_{\mu\lambda}a_{\lambda\nu}a_{\nu\mu} + (1/4)C\,(a_{\mu\nu})^2. \quad (1.37)$$

For the transition it has been used $A(T) = A_0(1-T^*/T)$. Here A_0, B, C (with $C < 2B^2/(9A_0)$) are positive dimensionless coefficients, and can be related to molecular quantities [45, 100–104]. The characteristic (pseudocritical) temperature T^* is also a model parameter. The value of A_0 depends on the proportionality coefficient chosen between $\boldsymbol{a}$ and $\overline{\langle \boldsymbol{uu} \rangle}$. The choice made in Eq. (1.13) implies $A_0 = 1$, cf. [45].

21

CHAPTER 1. NON-POLAR HARD-ROD FLUIDS

For lyotropic liquid crystals or rod dispersions, the concentration c of non spherical particles in a solvent rather than the temperature determines the phase transition, i.e., in this case one has $A \propto (1 - c/c^*)$, where c^* is a pseudo-critical concentration [103]. In Ref. [105], similar equations have been used to study the flow-alignment and rheology of semi-dilute polymer solutions, where c^* denotes the overlap concentration.

The Landau-de Gennes potential is related to the isotropic and uniaxial nematic equilibrium state. For biaxial nematic equilibrium a term proportional to $(a_{\sigma\rho}a_{\rho\kappa}a_{\sigma\kappa})^2$ has to be added [97, 106–108]. In the following the focus is on uniaxial nematic equilibrium. The Landau-de Gennes potential does not restrict the order parameter to physically admissible values. Later in section (1.6) an amended potential is introduced and discussed.

The Landau-de Gennes potential can be extended to the description of spatially inhomogeneous alignment by including gradient terms in the Landau-de Gennes potential [97, 109], for the lowest order

$$\Phi = \Phi^{LDG} + \frac{1}{2}L_1 \nabla_\lambda a_{\mu\nu} \nabla_\lambda a_{\mu\nu} + \frac{1}{2}L_2 \nabla_\lambda a_{\lambda\nu} \nabla_\mu a_{\mu\nu}. \tag{1.38}$$

The most general distorsion energy for uniaxial alignment ($\mathbf{a} \propto \overline{\mathbf{nn}}$) with respect to the head-tail symmetry is given by the Frank elastic free energy contribution [109]

$$F_d = \frac{1}{2}K_1 \left(\nabla \cdot \mathbf{n}\right)^2 + \frac{1}{2}K_2 \left(\mathbf{n} \cdot \nabla \times \mathbf{n}\right)^2 + \frac{1}{2}K_3 \left(\mathbf{n} \times \nabla \times \mathbf{n}\right)^2. \tag{1.39}$$

K_1, K_2 and K_3 are referred to as the Franks elastic constants for splay, twist and bend distorsions. The parameters L_1, L_2 are related to the curvature elastic constants K_1, K_2 and K_3 [109] via

$$K_1 = K_3 \propto (L_1 + \frac{1}{2}L_2)a_{\text{eq}}^2, \quad K_2 \propto L_1 a_{\text{eq}}^2. \tag{1.40}$$

To obtain the full anisotropy it is necessary to introduce an additional term to the Landau-de Gennes potential of the form

$$a_{\mu\nu} \left(\nabla_\lambda a_{\rho\mu}\right) \left(\nabla_\lambda a_{\rho\nu}\right). \tag{1.41}$$

Terms of this type arise in the calculation of the elasticity coefficients involving second and forth order tensors [110]. For simplicity it is assumed that all three coefficients are equal (one constant approximation), i.e. $L_2 = 0$. Frequently, ξ_a^2 as a characteristic molecular length scale is used for the coefficient L_1.

Note, in general the one constant approximation is not fulfilled for elongated particles in the spirit of mean-field theory [111]. However, for the mesoscopic theory presented here it is believed that it is acceptable for a first approximation.

1.3 Hydrodynamic Equations

1.3.1 Relaxation Equation for the Alignment Tensor

The isotropic-to nematic phase transition and equilibrium properties can be modeled within the Landau-de Gennes theory. However, to investigate the flow behavior a theoretical description of non-equilibrium states is needed. Non-equilibrium phenomena can be studied in

1.3. HYDRODYNAMIC EQUATIONS

the framework of irreversible thermodynamics or with dynamical equations for probability distribution functions (e.g. Fokker-Planck approach). The thermodynamical approach lacks of molecular details and therefore is more general. On the other hand the description of a specific material with known microscopic parameters is rather difficult. This thesis focus on general flow phenomena of anisotropic fluids and prefer the thermodynamic approach.

The starting point is the assumption that the generalized fundamental Gibbs relation [45]

$$\frac{ds}{dt} = T^{-1}\left(\frac{du}{dt} + p\frac{d\rho^{-1}}{dt}\right) - T^{-1}\frac{dg}{dt} \quad (1.42)$$

holds true for dynamic phenomena. The specific Gibbs free potential $g(\mathbf{a}, \nabla \mathbf{a})$ is associated with the alignment and the gradient of the alignment. A reasonable Ansatz for the Gibbs free potential is the Landau-de Gennes potential (1.38) except for proportionality. Based on the entropy production and a balance equation for the alignment tensor the relaxation equation for $\mathbf{a}$ in the presence of a flow field $\mathbf{v}$ yields [45, 47]

$$\begin{aligned}\frac{d}{dt}a_{\mu\nu} &- 2\overline{\varepsilon_{\mu\lambda\kappa}\omega_\lambda a_{\kappa\nu}} - 2\kappa_{\mathrm{a}}\overline{\Gamma_{\mu\lambda}a_{\lambda\nu}} = \\ &- \nabla_\lambda b_{\lambda\mu\nu} + \frac{\xi_{\mathrm{a}}^2}{\tau_{\mathrm{a}}}\triangle a_{\mu\nu} - \tau_{\mathrm{a}}^{-1}\Phi_{\mu\nu}^a(\mathbf{a}) - \sqrt{2}\,\frac{\tau_{\mathrm{ap}}}{\tau_{\mathrm{a}}}\Gamma_{\mu\nu},\end{aligned} \quad (1.43)$$

where the substantial time derivative is given by $\frac{d}{dt} = \partial_t + v_\lambda \nabla_\lambda$. The constants τ_a and τ_{ap} are phenomenological relaxation times with $\tau_a > 0$ and τ_{ap} having either sign. The parameter κ_{a} gives for the special values $\kappa_{\mathrm{a}} = 0$ the corotational and $\kappa_{\mathrm{a}} = 1$ the codeformational time derivative, respectively [22]. The parameters κ_{a}, τ_{a}, τ_{ap} can related to microscopic variables.

The symmetric traceless tensor

$$\Phi_{\mu\nu}^a(\mathbf{a}) \equiv \frac{\delta \Phi^{\mathrm{LDG}}}{\delta a_{\mu\nu}} \quad (1.44)$$

is the derivative of the Landau-de Gennes potential function Φ with respect to the alignment tensor. The tensors $\Gamma_{\mu\nu}$ and ω_λ denote the symmetric traceless part of the velocity gradient tensor (strain rate tensor) $\Gamma_{\mu\nu} \equiv \overline{\nabla_\mu v_\nu}$, and the averaged angular velocity ω_λ, respectively. The third rank tensor $b_{\lambda\mu\nu}$ is due to the tensor flux associated with the alignment and is given by

$$b_{\lambda\mu\nu} = -D_{\mathrm{a}}\nabla_\lambda\left(\Phi_{\mu\nu}^{\mathrm{a}} - \xi_{\mathrm{a}}\triangle a_{\mu\nu}\right). \quad (1.45)$$

1.3.2 Constitutive Equation for the Pressure Tensor

In the following the constitutive equation for the pressure tensor is presented. In Cartesian tensor notation the pressure tensor $P_{\nu\mu}$ occurring in the momentum balance equation (no external field, ρ is the mass density)

$$\rho\frac{dv_\mu}{dt} + \nabla_\nu P_{\nu\mu} = 0 \quad (1.46)$$

CHAPTER 1. NON-POLAR HARD-ROD FLUIDS

is decomposed according to

$$P_{\nu\mu} = P\delta_{\nu\mu} + \frac{1}{2}\varepsilon_{\nu\mu\lambda}p_\lambda + \overline{p}_{\nu\mu}. \tag{1.47}$$

Here $P = \frac{1}{3}P_{\lambda\lambda}$ is the trace part, $\overline{p_{\mu\nu}}$ is the symmetric traceless part of the tensor and $p_\lambda^a = \varepsilon_{\lambda\nu\mu}P_{\nu\mu}$ is the component of the pseudo vector associated with the antisymmetric part of the pressure tensor.

The trace part P is identified with the hydrostatic pressure linked with the local density and temperature by the equilibrium equation of state. The symmetric traceless friction pressure tensor consists of an 'isotropic' contribution as already present in fluids composed of spherical particles or in fluids of non-spherical particles in an perfectly 'isotropic state' with zero alignment, and a part explicitly depending on the alignment tensor:

$$\overline{p}_{\mu\nu} = -2\eta_{\text{iso}}\Gamma_{\mu\nu} + \overline{p_{\mu\nu}^{\text{al}}}, \tag{1.48}$$

with [47]

$$\overline{p_{\mu\nu}^{\text{al}}} = \frac{\rho}{m}k_BT\left(\sqrt{2}\frac{\tau_{\text{ap}}}{\tau_a}\Phi_{\mu\nu}^a - \sqrt{2}\frac{\tau_{\text{ap}}}{\tau_a}\xi_a^2\triangle a_{\mu\nu} - 2\kappa_a\overline{a_{\mu\lambda}\Phi_{\lambda\nu}^a} + 2\kappa_a\xi_a^2\overline{a_{\mu\lambda}\triangle a_{\lambda\nu}}\right)$$

for vanishing alignment tensor flux. Here m is mass of a particle, ρ/m is the number density, and $p_{\text{kin}} = \frac{\rho}{m}k_BT$ is the equilibrium kinetic pressure which is used as reference value for pressures.

In equilibrium one has $\Phi^a(\mathbf{a}) = \mathbf{0}$ and consequently $\overline{\mathbf{p_{\text{al}}}} = \mathbf{0}$. The occurrence of the same coupling coefficients τ_{ap} in (1.49) as in (1.43) is due to an Onsager symmetry relation [112]. For studies of the rheological properties in the isotropic and in the nematic phases with stationary flow alignment, following from (1.43) and (1.49), see [45, 47, 77, 113].

The conservation of the total angular momentum implies that the time change of the internal angular momentum is balanced by $\mathbf{p}^a$ in the absence of external torques. Due to $J_\mu = \theta\omega_\mu$, where ω_μ is the average angular velocity and θ a moment of inertia, and with the ansatz

$$p_\mu^a = \tau_r^{-1}(\omega_\mu - \frac{1}{2}\varepsilon_{\mu\lambda\rho}\partial_\lambda v_\rho). \tag{1.49}$$

one obtain

$$\frac{d}{dt}\omega_\mu = -\tau_r^{-1}(\omega_\mu - \frac{1}{2}\varepsilon_{\mu\lambda\rho}\partial_\lambda v_\rho), \tag{1.50}$$

The relaxation time τ_r measures how fast the average angular velocity follows the vorticity of the fluid. For dense fluids this relaxation time is rather small. This implies that the angular velocity is equal the vorticity, i.e.

$$\omega_\mu = \frac{1}{2}\varepsilon_{\mu\lambda\rho}\partial_\lambda v_\rho. \tag{1.51}$$

For the further discussion this assumption is used.

1.4. FLOW GEOMETRY

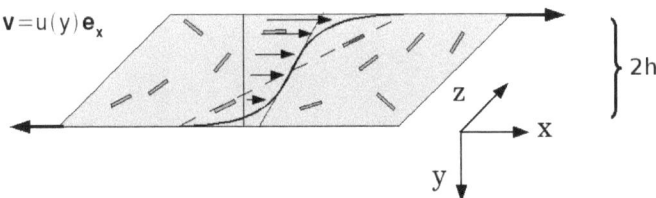

Figure 1.4: In the plane Couette flow geometry is displayed. The plates are infinitely long and lay in the xz-plane. The velocity profile is effectively one-dimensional.

1.4 Flow Geometry

In this thesis the simple Couette flow geometry is chosen for the investigation of the orientational behavior and flow properties of rod dispersions. In the plane Couette flow geometry (Couette cell) the fluid is between two plates. One plate is fixed at rest and the other moves with the speed u^w. The plates are infinitely long and lay in the xz-plane (see, Fig. 1.4). The geometry simplifies the system to efficiently 1-dimension. The velocity dependence is assumed to be $\mathbf{v}(y) = (u(y), 0, 0)^t$ and the alignment tensor to be $\mathbf{a} = \mathbf{a}(y)$. In that case the strain rate tensor and the vorticity are given by

$$\Gamma_{\mu\nu} = \begin{pmatrix} 0 & 0 & 0 \\ \frac{1}{2}\partial_y u(y) & 0 & 0 \\ 0 & 0 & 0 \end{pmatrix} \text{ and } \omega_\mu = \begin{pmatrix} 0 \\ 0 \\ -\frac{1}{2}\partial_y u(y) \end{pmatrix}, \text{ respectively.} \quad (1.52)$$

1.5 Scaled Variables

For numerical studies it is reasonable to scale the variables. Depending on the application different scalings are common. Here two are introduced, that differ on the time-scaling. For homogeneous systems the time is scaled according to a specific relaxation time. Otherwise, for heterogeneous systems the time is scaled by an effective shear rate.

1.5.1 Relaxation Time Scaling

The alignment tensor is expressed in units of the value of the order parameter at the isotropic-nematic phase transition, [45, 47, 113]

$$a^*_{\mu\nu} = \frac{a_{\mu\nu}}{a_K}, \quad a_K = \frac{2B}{3C} \quad (1.53)$$

25

CHAPTER 1. NON-POLAR HARD-ROD FLUIDS

occurring at the temperature $T_K > T^*$. With the reduced temperature variable

$$\vartheta \equiv \frac{9}{2}\frac{AC}{B^2} = \frac{1 - T^*/T}{1 - T^*/T_K} \tag{1.54}$$

the temperature dependence of the uniaxial equilibrium alignment is $a_{\text{eq}} = 0$ for $\vartheta \geq 9/8$ (isotropic phase) and

$$a_{\text{eq}}/a_K = \tfrac{1}{4}(3 + \sqrt{9 - 8\vartheta}), \quad \text{for } \vartheta < 9/8 \text{ (nematic phase)}. \tag{1.55}$$

Notice, that $\vartheta = 1$ corresponds to the equilibrium phase coexistence temperature, for vanishing coupling. The values $\vartheta = 9/8$ and $\vartheta = 0$ are the upper and lower limits of the metastable nematic and isotropic states, respectively. The quantity $\delta_K = 1 - T^*/T_K$ which sets a scale for the relative difference of the temperature from the pseudocritical temperature T^* to the temperature from equilibrium phase transition is known from experiments to be of the order 0.1 to 0.001. On the other hand, it is related to the coefficients occurring in the potential function according to

$$\delta_K = \frac{2}{9}\frac{B^2}{A_0 C} = \frac{1}{2}a_K^2 \frac{C}{A_0}. \tag{1.56}$$

The derivative Φ^a of the potential function in (1.43) can be written as

$$\Phi^a_{\mu\nu} = \Phi_{\text{ref}}\, \Phi^{a*}_{\mu\nu}(\mathbf{a}^*), \tag{1.57}$$

$$\Phi_{\text{ref}} = a_K \frac{2}{9}\frac{B^2}{C} = a_K \delta_K A_0, \quad a^*_{\mu\nu} = a_{\mu\nu}/a_K, \tag{1.58}$$

$$\Phi^{a*}_{\mu\nu}(\mathbf{a}) = \vartheta a^* - 3\sqrt{6 a^*_{\mu\lambda} a^*_{\lambda\nu}} + 2a^*_{\rho\sigma} a^*_{\rho\sigma} a^*_{\mu\nu}. \tag{1.59}$$

Clearly, the variable ϑ suffices to characterize the equilibrium behavior determined by $\Phi^a = \mathbf{0}$. The variable ϑ can also be interpreted as a density or concentration variable according to $\vartheta = (1 - c/c^*)/(1 - c_K/c^*)$ where c stands for the concentration (eg. lyotropic liquid crystals).

In the relaxation time scaling, times and shear rates are made dimensionless with a convenient reference time. The relaxation time of the alignment in the isotropic phase is $\tau_a A_0^{-1}(1 - T^*/T)^{-1}$ showing a pre-transitional increase. This relaxation time, at the coexistence temperature T_K, is used as a reference time

$$\tau_{\text{ref}} = \tau_a(1 - T^*/T_K)^{-1} A_0^{-1} = \tau_a \delta_K^{-1} A_0^{-1} = \tau_a \frac{9C}{2B2} = \tau_a\, a_K\, \Phi_{\text{ref}}^{-1}. \tag{1.60}$$

The shear rates are expressed in units of τ_{ref}^{-1}. For homogeneous systems, as this scaling is used, it is assumed that the shear rate $\dot\gamma$ is constant, i.e. $u(y) = \dot\gamma y$. The scaled shear rate, being a product of the true shear rate and the relevant relaxation time, is also referred to as 'Weissenberg-number' $Wi = \frac{\tau_a}{A_K}\dot\gamma$, where $A_K = \delta_K A_0$. Instead of the ratio τ_{ap}/τ_a, the tumbling parameter

$$\lambda_K = -(2/3)\sqrt{3}\,\frac{\tau_{\text{ap}}}{\tau_a}\, a_K^{-1} \tag{1.61}$$

1.5. SCALED VARIABLES

is used. The relaxation equation (Eq.1.43) for a spatially homeogenous alignment tensor in scaled variables yields

$$\frac{d}{dt^*}a^*_{\mu\nu} - 2\overline{\varepsilon_{\mu\lambda\rho}\omega^*_\lambda a^*_{\rho\nu}} - 2\kappa_a \overline{\Gamma^*_{\mu\lambda}a^*_{\lambda\nu}} = -\Phi^{a*}_{\mu\nu} + \sqrt{\frac{3}{2}}\lambda_K \Gamma^*_{\mu\nu}. \tag{1.62}$$

Here the dimensionless time $t^* = t\,\tau_{\text{ref}}^{-1}$, $\Gamma^*_{\mu\nu}$ and $\omega^*_{\mu\nu}$ as the symmetric traceless part of the dimensionaless velocity gradient $\nabla^*_\mu v^*_\nu$ and the scaled vorticity $\frac{1}{2}\varepsilon_{\lambda\mu\nu}\nabla^*_\mu v^*_\nu$ is used, respectively.

The flow gradient $\nabla^*_\mu v^*_\nu$ is equal to the dimensionless shear rate

$$\dot{\gamma}^* = \dot{\gamma}\tau_{\text{ref}} = Wi. \tag{1.63}$$

1.5.2 Shear Rate Scaling

For heterogeneous systems it is common to use a different scaling (see, [87]). A naturally time scale for the system is given by the effective shear rate $t_{\text{ref}}^{-1} = u^w/2h = \dot{\gamma}^{\text{eff}}$, where $2h$ denotes the plate separation of the Couette cell and u^w the velocity at the wall. The scaled variables reads ($p_{\text{kin}} = \frac{\rho}{m}k_B T$ denotes the kinetic pressure)

$$a^*_{\mu\nu} = \frac{a_{\mu\nu}}{a_K}, \quad v^*_\mu = \frac{v_\mu}{u^w}, \quad x^*_\mu = \frac{x_\mu}{2h}, \quad p^* = \frac{p}{p_{\text{kin}}}, \quad t^* = \frac{t}{t_{\text{ref}}}. \tag{1.64}$$

The scaled form of the relaxation equation (1.43) is given by

$$\frac{d}{dt^*}a^*_{\mu\nu} - 2\overline{\varepsilon_{\mu\lambda\rho}\omega^*_\lambda a^*_{\rho\nu}} - 2\kappa_a \overline{\Gamma^*_{\mu\lambda}a^*_{\lambda\nu}} = \tag{1.65}$$
$$+ \bar{D}_a \Delta^* \Phi^{a*}_{\mu\nu} - \frac{\bar{D}_a Wi}{Er}\Delta^{*2} a^*_{\mu\nu} + \frac{1}{Er}\Delta a_{\mu\nu} - \frac{1}{Wi}\Phi^{a*}_{\mu\nu} + \sqrt{\frac{3}{2}}\lambda_K \Gamma^*_{\mu\nu},$$

with the derivative of the potential function $\Phi^{a*}_{\mu\nu}$. The Weissenberg number Wi ($\bar{D}_R$ is the averaged rotational diffusion constant) and the Ericksen number are given by

$$Wi = \frac{\tau_a \dot{\gamma}^{\text{eff}}}{A_K} = \frac{\dot{\gamma}^{\text{eff}}}{6\bar{D}_R A_K}, \quad Er = Wi\left(\frac{2h}{\xi_0}\right)^2 A_K. \tag{1.66}$$

The scaled diffusion constant reads

$$\bar{D}_a = \frac{a_K A_K}{2hu^w}D_a. \tag{1.67}$$

For heterogeneous systems the parameter Wi and Er plays a significant role in the formation of structure. The Weissenberg number expresses the competition between flow induced distorsion and molecular relaxation. The Ericksen number is a measure for the ratio of the viscous torque due to the flow and Franks elastic distorsions.

The constitutive equation and the momentum equation in scaled variables are

$$P^*_{\mu\nu} = p^*\delta_{\mu\nu} - 2\nu_{iso}\Gamma^*_{\mu\nu} - \sqrt{\frac{3}{2}}\lambda_K \Phi^{a*}_{\mu\nu} + \sqrt{\frac{3}{2}}\frac{Wi}{Er}\Delta a^*_{\mu\nu} \tag{1.68}$$
$$- 2\kappa_a \overline{a^*_{\mu\lambda}\Phi^{a*}_{\lambda\nu}} + \kappa_a \frac{2Wi}{Er}\overline{a^*_{\mu\lambda}\Delta a^*_{\lambda\nu}},$$

$$\frac{d}{dt^*}v^*_\mu = -\frac{1}{\beta}\nabla_\lambda \left(\iota_K p^*\delta_{\lambda\mu} + \overline{P^*_{\lambda\mu}}\right) = \frac{1}{\beta}\nabla_\lambda(-\iota_K p^*\delta_{\lambda\mu} + \tau_{\lambda\mu}). \tag{1.69}$$

CHAPTER 1. NON-POLAR HARD-ROD FLUIDS

variables	relaxation time scaling	shear rate scaling
t	$t^* = \frac{A_K}{\tau_a} t$	$t^* = \frac{u^w}{2h} t$
$\mathbf{v}$	$\mathbf{v}^* = \frac{2h}{\tau_a A_K} \mathbf{v}$	$\mathbf{v}^* = \frac{\mathbf{v}}{u^w}$
x		$x^* = \frac{x}{2h}$
$\mathbf{a}$		$\mathbf{a}^* = \frac{\mathbf{a}}{a_K}$
p		$p^* = \frac{p}{p_{\text{kin}}}$
parameters		
Wi	$\frac{\tau_a}{A_K} \dot{\gamma}$	$\frac{\tau_a}{A_K} \dot{\gamma}^{\text{eff}}$
Er	-	$\tau_a \dot{\gamma}^{\text{eff}} \left(\frac{2h}{\xi_0}\right)^2$
λ_K	$-\frac{2}{3} \frac{\tau_{ap}}{\tau_a a_K}$	
ν_{iso}	-	$\frac{\eta_{\text{iso}}}{p_{\text{kin}} a_K^2 A_K} \dot{\gamma}^{\text{eff}}$
β	-	$\frac{\rho (u^w)^2}{p_{\text{kin}} a_K^2 A_K}$

Table 1.1: The table shows the similarities and differences of the relaxation time scaling (homogeneous alignment tensor) compared to the effective shear rate scaling (heterogeneous alignment tensor).

The parameter β measures the strength of the inertia related to viscosity forces and the coefficient ν_{iso} is related to the second Newtonian viscosity η_{iso}, viz

$$\beta = \frac{\rho (u^w)^2}{p_{\text{kin}} a_K^2 A_K}, \quad \nu_{\text{iso}} = \frac{\eta_{\text{iso}}}{p_{\text{kin}} a_K^2 A_K} \dot{\gamma}^{\text{eff}}. \tag{1.70}$$

The parameter ι_K corresponds to $\iota_K = (a_K A_K)^{-1}$.

1.6 Amended Landau-de Gennes Potential

The Landau-de Gennes potential does not restrict the order parameter be within its physically imposed bounds. For numerical studies and for elongational flows (as it is shown in the next part) it is necessary to restrict the alignment tensor such that its magnitude is bounded. In the scaled formulation the expansion of the new potential in terms of the alignment should reduce to the Landau-de Gennes expression (1.59) when terms of higher than 4th order are disregarded. Thus the ansatz

$$\Phi = (1/2)\, \vartheta\, a_{\mu\nu} a_{\mu\nu} - \sqrt{6}\, (a_{\mu\lambda} a_{\lambda\nu}) a_{\mu\nu} + \varphi \tag{1.71}$$

is made where φ should reduce to $(1/2)(a_{\mu\nu} a_{\mu\nu})^2$ for small values of the alignment. As mentioned, it is understood that $\mathbf{a}$ and Φ stand for $\mathbf{a}^*$ and Φ^*. A simple choice for φ which

1.6. AMENDED LANDAU-DE GENNES POTENTIAL

ensures that the magnitude of the alignment does not exceed $a_{\max}$ is

$$\varphi = -(1/2)\, a_{\max}^4 \ln\left(1 - \frac{(a_{\mu\nu}a_{\mu\nu})^2}{a_{\max}^4}\right). \tag{1.72}$$

In the case of a uniaxial alignment where one has $a_{\mu\nu} = a(3/2)^{1/2}\,\overline{n_\mu n_\nu}$, the potential function reduces to a function of the scalar order parameter a, viz.:

$$\Phi = (1/2)\,\vartheta\, a^2 - a^3 - (1/2)\, a_{\max}^4 \ln\left(1 - \frac{a^4}{a_{\max}^4}\right). \tag{1.73}$$

In the following, $a_{\max} = 2.5$ is chosen. This is a plausible value for thermotropic liquid crystals, where the Maier-Saupe order parameter $S = \langle P_2 \rangle$ is about 0.4 at the transition temperature. Thus the maximum value 1 for S is larger by the factor 2.5. Fig. 1.5 show the Landau-de Gennes and the amended potential for $a_{\max} = 2.5$. The differences are very small. In the Landau-de Gennes case one has $a = a_K = 1$ at the transition temperature $\vartheta = \vartheta_K = 1$. For the amended potential with $a_{\max} = 2.5$ one has the transition at $\vartheta = \vartheta_K \approx 0.9883$ with $a_K \approx 0.9667$. Due to the small difference between these values it convenient to maintain the Landau-de Gennes scaling for the physical variables.

1.6.1 Theoretical Motivation

To justify the educated guess of the amended potential (1.71) it will be shown that within the Fokker-Planck description of the orientational distribution function one can derive a Landau-de Gennes typ potential that naturally exhibit a restriction of the order parameter. Based on Onsager's excluded volume model of hard rods, the first order corrections to the Maier Saupes mean field potential were calculated in [114]. However, the second and higher order corrections lead to a restriction of the order parameter.

Starting with the generalized Fokker-Planck equation for the probability distribution function $\rho^{\mathrm{or}}(\mathbf{u},t)$ in the presence of a flow field as was given independently by Hess and Doi [100, 104], viz.

$$\partial_t \rho^{\mathrm{or}} = -\mathcal{L}_\lambda [\varepsilon_{\lambda\mu\nu} u_\mu (k_{\nu\sigma} u_\sigma \rho^{\mathrm{or}})] + \mathcal{L}_\lambda \bar{D}_{\mathrm{R}} \rho^{\mathrm{or}} \mathcal{L}_\lambda \left(\frac{\delta A}{\delta \rho^{\mathrm{or}}(\mathbf{u})}\right). \tag{1.74}$$

Here, $\mathcal{L}_\lambda = \varepsilon_{\lambda\mu\nu} u_\mu \frac{\partial}{\partial u_\nu}$ is the rotational operator, $\frac{\partial}{\partial u_\nu}$ the derivative on the unit sphere, $k_{\mu\nu} = \partial_\mu v_\nu$ the velocity gradient, $\bar{D}_{\mathrm{R}}$ the average rotational diffusion constant and $\frac{\delta A}{\delta f}$ the functional derivative of $A = A_0 + A_1$, the free energy per molecule modulo $k_B T$. The free energy consists of the loss of entropy with molecular alignment

$$A_0 = \ln\nu - 1 + \langle \ln\rho^{\mathrm{or}}(\mathbf{u}) \rangle \tag{1.75}$$

and the Onsager free energy of steric interaction in the second virial approximation

$$A_1 = \frac{U}{2} \langle\langle \sqrt{1 - (u_\lambda w_\lambda)^2} \rangle\rangle, \tag{1.76}$$

CHAPTER 1. NON-POLAR HARD-ROD FLUIDS

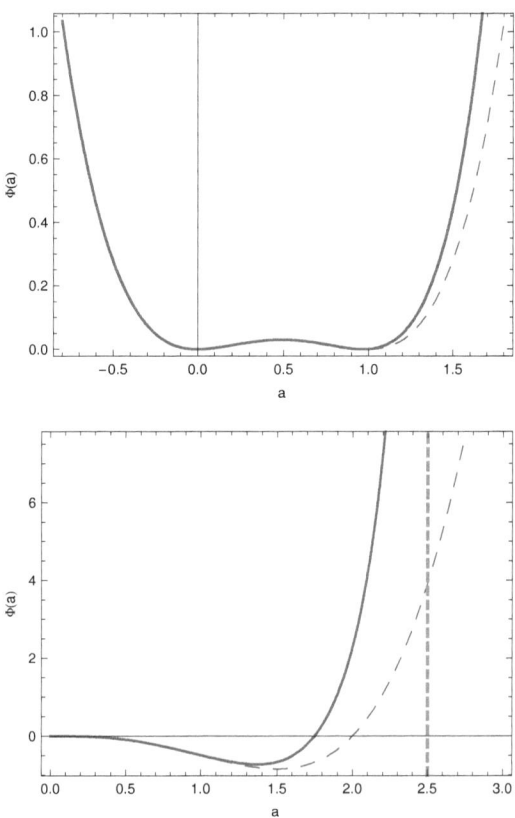

Figure 1.5: The Landau-de Gennes potential (dashed line) and the amended potential (full line) as a function of the scalar order parameter a for $\vartheta = 0$ (lower) and $\vartheta = 1$ (upper).

1.6. AMENDED LANDAU-DE GENNES POTENTIAL

where $U = 2bL^2\nu$ is the reduced excluded-volume, $2b$ and L are the diameter and the length of the rodlike molecules, and ν is the number of molecules per unit volume. Here and below, the following notation for averages of arbitrary functions $F(\mathbf{u})$ is used:

$$\langle F(\mathbf{u})\rangle = \int_{S^2} F(\mathbf{u})\rho^{\mathrm{or}}(\mathbf{u})d^2u, \quad \langle\langle F(\mathbf{u},\mathbf{w})\rangle\rangle = \int_{S^2}\int_{S^2} F(\mathbf{u},\mathbf{w})\rho^{\mathrm{or}}(\mathbf{u})\rho^{\mathrm{or}}(\mathbf{w})\, d^2u\, d^2w \quad (1.77)$$

In principle, a hierarchy of moment equations can be derived from the Fokker-Planck equation (1.74). However, due to the nonlinearity of A_1, the time evolution equation of the alignment tensor $\mathbf{a}$ couples directly to all higher order moments, which makes further analytical studies impractical. In [114], systematic approximations to the functional A_1 have been proposed that lead to simpler hierarchies of moment equations which can further be analyzed. The first and second terms in the approximation $A_1 \approx A_1^{(1)} + A_1^{(2)}$ are [114]

$$A_1^{(1)} = \frac{U}{2}\sqrt{1 - \langle u_\mu u_\nu\rangle\langle u_\mu u_\nu\rangle} \quad (1.78)$$

$$A_1^{(2)} = -\frac{U}{16}\langle\langle [(u_\lambda w_\lambda)^2 - \langle u_\mu u_\nu\rangle\langle u_\mu u_\nu\rangle]^2\rangle\rangle(1 - \langle u_\mu u_\nu\rangle\langle u_\mu u_\nu\rangle)^{-\frac{3}{2}}. \quad (1.79)$$

The functional derivative of $A_1^{(1)}$ and $A_1^{(2)}$ are derived as

$$\frac{\delta}{\delta f}A_1^{(1)} = \frac{U u_\mu u_\nu \langle u_\mu u_\nu\rangle}{2\sqrt{1 - \langle u_\kappa u_\sigma\rangle\langle u_\kappa u_\sigma\rangle}} \quad (1.80)$$

$$\frac{\delta}{\delta f}A_1^{(2)} = -\frac{U}{8}\frac{u_\mu u_\nu u_\kappa u_\sigma \langle u_\mu u_\nu u_\kappa u_\sigma\rangle - 2(u_\mu u_\nu\langle u_\mu u_\nu\rangle)(\langle u_\kappa u_\sigma\rangle\langle u_\kappa u_\sigma\rangle)}{(1 - \langle u_\delta u_\xi\rangle : \langle u_\delta u_\xi\rangle)^{\frac{3}{2}}}. \quad (1.81)$$

By Prager's procedure, the relaxation of the alignment tensor $a_{\mu\nu}$ can be derived form Eq. (1.74, 1.78) and (1.79), viz.

$$\partial_t a_{\mu\nu} = \bar{D}_r(\langle\mathcal{L}_\lambda\mathcal{L}_\lambda t_{\mu\nu}\rangle) - \frac{U}{2\sqrt{1 - \langle u_\delta u_\xi\rangle\langle u_\delta u_\xi\rangle}}\langle\mathcal{L}_\lambda(t_{\mu\nu})\mathcal{L}_\lambda u_\kappa u_\sigma \langle u_\kappa u_\sigma\rangle\rangle)$$

$$+ D_r \frac{U}{\sqrt{(1 - \langle u_\delta u_\xi\rangle\langle u_\delta u_\xi\rangle)^3}}\left(\frac{1}{4}\langle(\mathcal{L}_\lambda t_{\mu\nu})\mathcal{L}_\lambda(u_\kappa u_\sigma \langle u_\kappa u_\sigma\rangle)\rangle\right. \quad (1.82)$$

$$\left. + \frac{1}{8}\langle(\mathcal{L}_\lambda t_{\mu\nu})\mathcal{L}_\lambda(u_\kappa u_\sigma u_\alpha u_\beta \langle u_\kappa u_\sigma u_\alpha u_\beta\rangle)\rangle\right),$$

where $a_{\mu\nu} = \langle t_{\mu\nu}\rangle$ and $t_{\mu\nu} = \overline{u_\mu u_\nu}$. The decoupling approximations $a_{\lambda\kappa}\langle u_\lambda u_\kappa u_\mu u_\nu\rangle = a_{\lambda\kappa}\langle u_\lambda u_\kappa\rangle\langle u_\mu u_\nu\rangle$, $\langle u_\mu u_\nu u_\lambda u_\kappa\rangle\langle u_\mu u_\nu u_\lambda u_\kappa\rangle = \langle u_\mu u_\nu\rangle\langle u_\lambda u_\kappa\rangle\langle u_\mu u_\nu u_\lambda u_\kappa\rangle$ is used and the uniaxial case $a_{\mu\nu} = q'\overline{n_\mu n_\nu}$ is considered, where $q' = \sqrt{\frac{3}{2}}a$. The relaxation equation for the scalar order parameter q' is derived as

CHAPTER 1. NON-POLAR HARD-ROD FLUIDS

$$\begin{aligned}
\partial_t q' &= -6 D_r \frac{\partial \phi(q', U)}{\partial S}, \\
\phi(q', U) &= \frac{q'^2}{2} - \frac{U'}{6}\sqrt{1-q'^2}\left(1 - \frac{3q}{2} + 2q'^2\right) - \frac{U'}{4}\arcsin(q') + \frac{U'}{6} \\
&\quad + \frac{U'}{\sqrt{(1-q'^2)^3}}\left(-\frac{1}{4}q'^7 + \frac{1}{12}q'^6 - \frac{3}{16}q'^5 + \frac{1}{4}q'^4 + \frac{21}{16}q'^3 - \frac{15}{16}q'^2 - \frac{7}{8}q'\right. \\
&\quad \left. + \frac{29}{48}[1 - \sqrt{(1-q'^2)^3}]\right) + \frac{7}{8}\arcsin(q'), \quad (1.83)
\end{aligned}$$

where $U' = \sqrt{3/2}\,U$. The integration constant is determined by the requirement $\Phi(0) = 0$. In addition to the first corrections calculated in [114], the second correction terms are singular for $q' \to \pm 1$. Hence, the use of approximations to Onsager's excluded volume potential leads to a restriction of the order parameter values in a natural way. It is interesting to note, that taking into account higher order corrections does not change the singularity since these terms produce higher order derivatives of $\sqrt{1 - u_\lambda w_\lambda}$. Note also, that although the use of different decoupling schemes lead to different forms of the potential (1.83), the singularity for $q' \to \pm 1$ remains unchanged. The order parameter q' is related to the Maier-Saupe order parameter by $q' = \sqrt{\frac{15}{2}} S$. The full tensorial form of (1.83) is difficult to receive and hence for further analyses the simple potential (1.71) with $a_{\max} = 2.5$ is used.

1.7 Component Form of the Model Equations

For the numerical analysis it is necessary to express the tensorial equations in component form. Using the basis tensors (1.32), one obtains from Eq. (1.62) for a plane Couette flow a system of coupled partial differential equations (in the shear rate scaling):

$$\begin{aligned}
\partial_t a_0 &= -\frac{1}{Wi}\Phi_0^a - \frac{1}{3}\sqrt{3}\,\kappa_a\, a_2\, \partial_y u + \bar{D}_a \partial_y^2 \Phi_0^a + \frac{1}{Er}\partial_y^2 a_0 - \frac{\bar{D}_a}{Er}\partial_y^4 a_0 \\
\partial_t a_1 &= -\frac{1}{Wi}\Phi_1^a + a_2 \partial_y u + \bar{D}_a \partial_y^2 \Phi_1^a + \frac{1}{Er}\partial_y^2 a_1 - \frac{\bar{D}_a}{Er}\partial_y^4 a_1, \quad (1.84) \\
\partial_t a_2 &= -\frac{1}{Wi}\Phi_2^a - a_1 \partial_y u + \frac{\sqrt{3}}{2}\lambda_K\,\partial_y u - \frac{1}{3}\sqrt{3}\,\kappa_a\, a_0\, \partial_y u + \bar{D}_a \Phi_2^a + \frac{1}{Er}\partial_y^2 a_2 - \frac{\bar{D}_a}{Er}\partial_y^4 a_2 \\
\partial_t a_3 &= -\frac{1}{Wi}\Phi_3^a + \frac{1}{2}(\kappa_a + 1)\,a_4 \partial_y u + \bar{D}_a \partial_y^2 \Phi_3^a + \frac{1}{Er}\partial_y^2 a_3 - \frac{\bar{D}_a}{Er}\partial_y^4 a_3, \\
\partial_t a_4 &= -\frac{1}{Wi}\Phi_4^a + \frac{1}{2}(\kappa_a - 1)\,a_3 \partial_y u + \bar{D}_a \partial_y^2 \Phi_4^a + \frac{1}{Er}\partial_y^2 a_4 - \frac{\bar{D}_a}{Er}\partial_y^4 a_4,
\end{aligned}$$

1.8. FURTHER MODELS AND APPROACHES

where $\Phi_i^a \equiv \boldsymbol{\Phi}^a : \mathbf{T}^i$ is given by

$$\begin{aligned}
\Phi_0^a &= (\vartheta - 3a_0 + 2a^2\psi)\,a_0 + 3(a_1^2 + a_2^2) - \frac{3}{2}(a_3^2 + a_4^2), \\
\Phi_1^a &= (\vartheta + 6a_0 + 2a^2\psi)\,a_1 - \frac{3}{2}\sqrt{3}(a_3^2 - a_4^2), \\
\Phi_2^a &= (\vartheta + 6a_0 + 2a^2\psi)\,a_2 - 3\sqrt{3}\,a_3 a_4, \\
\Phi_3^a &= (\vartheta - 3a_0 + 2a^2\psi)\,a_3 - 3\sqrt{3}(a_1 a_3 + a_2 a_4), \\
\Phi_4^a &= (\vartheta - 3a_0 + 2a^2\psi)\,a_4 - 3\sqrt{3}(a_2 a_3 - a_1 a_4).
\end{aligned} \qquad (1.85)$$

The notation $a^2 \equiv a_0^2 + a_1^2 + a_2^2 + a_3^2 + a_4^2$ is used. The quantity ψ is equal to 1 for the Landau-de Gennes potential and

$$\psi = \left(1 - \frac{(a^2)^2}{a_{\max}^4}\right)^{-1} \qquad (1.86)$$

for the amended potential function (1.71). The parameters $\vartheta, \lambda_K, \kappa$ were introduced in the foregoing section. The momentum equation on component form yields for vanishing alignment tensor flux

$$\begin{aligned}
\frac{\partial u}{\partial t} =\ & \frac{\nu_{\text{iso}}}{\beta}\partial_y^2 u + \sqrt{\frac{3}{2}}\frac{\lambda_K}{\beta}\partial_y \Phi_2^a - \sqrt{\frac{3}{2}}\frac{\lambda_K}{\beta}\frac{Wi}{Er}\partial_y^3 a_2 \\
& + \frac{\kappa_a}{\beta}\partial_y\left(\frac{1}{2\sqrt{2}}a_3\Phi_4^a - \frac{1}{2\sqrt{2}}\frac{Wi}{Er}a_3\partial_y^2 a_4 - \frac{1}{\sqrt{6}}a_0\Phi_2^a + \frac{1}{\sqrt{6}}\frac{Wi}{Er}a_0\partial_y^2 a_2 \right. \\
& \left. + \frac{1}{2\sqrt{2}}a_4\Phi_3^a - \frac{1}{2\sqrt{2}}\frac{Wi}{Er}a_4\partial_y^2 a_3 - \frac{1}{\sqrt{6}}a_2\Phi_0^a - \frac{1}{\sqrt{6}}\frac{Wi}{Er}a_2\partial_y^2 a_0\right).
\end{aligned} \qquad (1.87)$$

For the investigation of the non-Newtonian behavior the first and second normal stress differences are useful. From equations (1.69) and

$$\tau_{\mu\nu} = \nu_{\text{iso}}\Gamma_{\mu\nu} + \sigma_{\mu\nu} \qquad (1.88)$$

one deduces expressions for the (dimensionless) shear stress σ_{xy}, and the normal stress differences $N_1 = \sigma_{xx} - \sigma_{yy}$ and $N_2 = \sigma_{yy} - \sigma_{zz}$ in terms of the dimensionless tensor components $\sigma_i \equiv \sigma_{\lambda\kappa}^{\text{al}} T_{\lambda\kappa}^i$. These relations are

$$\tau_{xy} = \nu_{\text{iso}}\partial_y u + \sigma_2, \quad N_1 = 2\sigma_1, \quad N_2 = -\sqrt{3}\sigma_0 - \sigma_1. \qquad (1.89)$$

1.8 Further Models and Approaches

Ericksen-Leslie Theory

Based on general conservation laws and constitutive equations Ericksen and Leslie derived a continuum theory for nematic liquid crystals [115, 116]. These equations are widely used. Here a brief review is given following [7, 117] and [44].

CHAPTER 1. NON-POLAR HARD-ROD FLUIDS

The stess tensor $\tau_{\mu\nu}$ in the linear momentum equation

$$\rho \frac{d}{dt} v_\mu = \nabla_\lambda \tau_{\lambda\mu} \tag{1.90}$$

can be split into a viscous part $\tau^{\rm v}_{\mu\nu}$, a elastic part $\tau^{\rm e}_{\mu\nu}$ and the isotropic hydrostatic stress p,

$$\tau_{\mu\nu} = -p\delta_{\mu\nu} + \tau^{\rm v}_{\mu\nu} + \tau^{\rm e}_{\mu\nu}. \tag{1.91}$$

The elastic and viscous stress is due to the orientational friction contribution. With the director n_μ and the vector $N_\mu = \dot{n}_\mu - \varepsilon_{\mu\rho\sigma}\omega_\rho n_\sigma$ (representing the change of the director with respect to the background fluid) the viscous stress yields

$$\tau^{\rm v}_{\alpha\beta} = \alpha_1 n_\alpha n_\beta n_\rho \Gamma_{\mu\rho} + \alpha_2 n_\alpha N_\beta + \alpha_3 n_\beta N_\alpha + \alpha_4 \Gamma_{\alpha\beta} + \alpha_5 n_\alpha n_\mu \Gamma_{\mu\beta} + \alpha_6 n_\beta n_\mu \Gamma_{\mu\alpha}. \tag{1.92}$$

Here $\Gamma_{\mu\nu} = \frac{1}{2}(\nabla_\nu v_\mu + \nabla_\mu v_\nu)$ is the rate of strain tensor and $\omega_\mu = \frac{1}{2}\varepsilon_{\mu\rho\sigma}\nabla_\rho v_\sigma$ the vorticity. The Leslie viscosity coefficients $\alpha_1...\alpha_6$ are not independent. In the framework of irreversible thermodynamics Onsager relation between the αs yields the Parodi relation [118]

$$\alpha_6 - \alpha_5 = \alpha_2 + \alpha_3. \tag{1.93}$$

The elastic part of the stress tensor reads

$$\sigma_{\beta\gamma} = -\frac{\delta F^{\rm d}}{\delta(\partial_\beta n_\gamma)} \partial_\alpha n_\gamma. \tag{1.94}$$

Here $F^{\rm d}$ is the distorsion part of the free energy functional. In general, the viscous part and the elastic part, respectively is not symmetric and yields an viscous torque $\boldsymbol{T}$ acting on the director [7]

$$T^{\rm e}_\mu = \varepsilon_{\mu\nu\lambda} n_\nu \frac{\delta F^{\rm d}}{\delta n_\lambda} \tag{1.95}$$

$$T^{\rm v}_\mu = (-\varepsilon_{\mu\nu\lambda} n_\nu \gamma_1 N_\lambda - \gamma_2 \Gamma_{\mu\lambda} n_\lambda), \tag{1.96}$$

where $\gamma_1 = \alpha_3 - \alpha_2$ and $\gamma_2 = \alpha_2 + \alpha_3$. The balance of torques $\boldsymbol{T}^{\rm e} + \boldsymbol{T}^{\rm v} = 0$ implies the equation of motion for the director $\boldsymbol{n}$, i.e.

$$\varepsilon_{\mu\lambda\nu} n_\lambda \left(\frac{dn_\nu}{dt} - \varepsilon_{\nu\rho\sigma}\omega_\rho n_\sigma - \lambda \Gamma_{\nu\rho} n_\rho + \frac{\delta F}{\delta n_\nu} \right) = 0, \tag{1.97}$$

with the "tumbling coefficient" $\lambda = -\frac{\gamma_2}{\gamma_1}$.

The Ericksen-Leslie theory can be derived from the tensorial theory when one assumes that the alignment is uniaxial and that the order parameter is constant. Then the parameters of the Ericksen-Leslie theory can be expressed in terms of the parameters governing the more general tensorial approach.

As it was shown previously for nematics, [47, 100, 113, 115], the relaxation times $\tau_{\rm a}$ and $\tau_{\rm ap}$ are proportional to the viscosity coefficients γ_1 and γ_2, i.e.

$$\gamma_1 = 3\frac{\rho}{m} k_{\rm B} T a_{\rm eq}^2 \tau_{\rm a}, \quad \gamma_2 = \frac{\rho}{m} k_{\rm B} T \left(2\sqrt{3} a_{\rm eq} \tau_{\rm ap} - \kappa_{\rm a} a_{\rm eq}^2 \right). \tag{1.98}$$

1.8. FURTHER MODELS AND APPROACHES

In [119] the temperature dependence of the tumbling coefficient λ is discussed.

The relaxation equation for the alignment tensor (1.65) for uniaxial distributions in scaled variables [44] reads

$$\frac{d}{dt}a = \beta(a)\Gamma_{\mu\nu}\overline{n_\mu n_\nu} - \Phi'(a) \tag{1.99}$$

$$\frac{d}{dt}n_\mu = \varepsilon_{\mu\nu\rho}\omega_\nu n_\rho + \lambda(a)\left(\Gamma_{\mu\rho}n_\rho - \Gamma_{\kappa\rho}n_\kappa n_\rho n_\mu\right), \tag{1.100}$$

with the abbreviations

$$\beta(a) = \kappa_a a + \frac{3}{2}\lambda_K, \quad \lambda(a) = \frac{\kappa_a}{3} + \frac{\lambda_K}{a} \tag{1.101}$$

and the derivative of the Landau-de Gennes potential for uniaxial alignment

$$\Phi'(a) = \theta a - 3a^2 + 2a^3. \tag{1.102}$$

In the limit of low shear rates ($\dot{\gamma} \ll 1$) the order parameter can assumed to be constant $a \approx a_{\text{eq}}$. In this case the dynamical equation for the director (1.100) reduce to the Ericksen Leslie equation (1.97). In this limit the tumbling coefficient $\lambda = -\gamma_2/\gamma_1 = \lambda(a_{\text{eq}}) = \lambda_{\text{eq}}$ is given by

$$\lambda_{\text{eq}} = \lambda_K \frac{a_K}{a_{\text{eq}}} + \frac{1}{3}\kappa_a, \tag{1.103}$$

where a_{eq} is recalled as the equilibrium value of the alignment in the nematic phase. Thus λ_{eq} is equal to λ_K at the transition temperature, corresponding to $\vartheta_{\text{eff}} = 1$, provided that $\kappa = 0$. Notice that λ_{eq}, in contradistinction to λ_K, is defined in the nematic phase only. In the limit of small shear rates $\dot{\gamma}$, the tumbling parameter is related to the Jeffrey tumbling period [120], see also [77]. Within the Ericksen-Leslie description, the flow alignment angle χ in the nematic phase is determined by

$$\cos(2\chi) = -\gamma_1/\gamma_2 = 1/\lambda_{\text{eq}}. \tag{1.104}$$

A stable flow alignment, at small shear rates, exists for $|\lambda_{\text{eq}}| > 1$ only. For $|\lambda_{\text{eq}}| < 1$ tumbling and an even more complex time-dependent behavior of the orientation occur. The quantity $|\lambda_{\text{eq}}| - 1$ can change sign as function of the variable ϑ. For $|\lambda_{\text{eq}}| < 1$ and in the limit of small shear rates $\dot{\gamma}$, the Jeffrey tumbling period [120] is related to the Ericksen-Leslie tumbling parameter λ_{eq} by $P_J = \frac{4\pi}{\dot{\gamma}\sqrt{1-\lambda_{\text{eq}}^2}}$, for a full rotation of the director.

In the following, λ_K, and κ_a, are considered as model parameters. The first one is essential for the coupling between the alignment and the viscous flow. The coefficients κ_a influences the orientational behavior quantitatively but do not seem to affect it in a qualitative way. If one wants to correlate the present theory with the flow behavior of the alignment in the isotropic phase, on the one hand, and in the nematic phase, on the other hand, for small shear rates where the magnitude of the order parameters is practically not altered, it suffices to study the case $\lambda_K \neq 0$, $\kappa = 0$, in order to match an experimental value of λ by the expression (1.103). Mesoscopic theories [100, 104, 121, 122] indicate that $\kappa \sim \lambda_K$.

CHAPTER 1. NON-POLAR HARD-ROD FLUIDS

Also the relations for the Leslie viscosity coefficients in the nematic phase are derived with the constitutive equation for the pressure tensor (1.48) exhibiting four viscosity coefficients [47]

$$\eta = \frac{\rho}{m} k_B T (\tau_p + \frac{1}{6}\kappa^2 a_{eq}^2 \tau_a) \qquad (1.105)$$

$$\eta_1 = \frac{\rho}{m} k_B T \kappa a_{eq}(-2\sqrt{3}\tau_{pa} - \frac{1}{2}\kappa a_{eq}\tau_a) \qquad (1.106)$$

$$\eta_2 = \frac{\rho}{m} k_B T a_{eq}(\sqrt{3}\tau_{pa} - \frac{1}{2}\kappa a_{eq}\tau_p) \qquad (1.107)$$

$$\eta_3 = \frac{\rho}{m} k_B T \frac{1}{2}\kappa^2 a_{eq}^2 \tau_a. \qquad (1.108)$$

The viscosities are related to the Leslie coefficients, viz

$$\eta = \frac{1}{2}\alpha_4 + \frac{1}{6}(\alpha_5 + \alpha_6), \ \eta_1 = \frac{1}{2}(\alpha_5 + \alpha_6), \qquad (1.109)$$

$$\eta_2 = \frac{1}{2}(\alpha_2 + \alpha_3), \ \eta_3 = \frac{1}{2}\alpha_1 \qquad (1.110)$$

The Leslie viscosity coefficients $\alpha_{1...6}$ as well as the viscosities η_i are not measurable in experiments. For a plane Couette flow geometry Miesowicz viscosities are the relevant viscosities. However, Miesowicz viscosities are directly related to the viscosities (1.105-1.108) by linear combinations (see [47]).

To summarize, the Ericksen-Leslie theory follows from the alignment tensor approach when the alignment tensor $\mathbf{a} \neq 0$ is uniaxial and when the effect of the shear flow on the magnitude of the order parameter can be disregarded. Then it suffices to use a dynamic equation for the 'director' $\mathbf{n}$ which is a unit vector parallel to the principal axis of the alignment tensor associated with its largest eigenvalue. This is a good approximation deep in the nematic phase and for small shear rates. For intermediate and large shear rates, the description of defects and, in particular, in the vicinity of the isotropic-nematic phase transition, the tensorial description is needed.

2
Polar Hard-Rod Fluids

2.1 Orientational Distribution and its Tensorial Representation

In the previous chapter uniaxially shaped particles with a collective behavior leading to the "head-tail" symmetry of the orientational distribution function at every time and space point were considered. Here it is assumed that each particle possesses a permanent electric or magnetic dipole moment characterized by the dimensionless unit vector $\mathbf{e}_i$, which encloses a fixed angle $\alpha_{\rm dip}$ with the particle axis such that $\mathbf{u}_i \cdot \mathbf{e}_i = \cos \alpha_{\rm dip}$ is independent of i, Fig. 2.1. In this case the "head-tail" symmetry of the orientational distribution function can be broken. The average orientation of the molecular axis $\mathbf{u}$ as in the previous chapter is described by the second rank alignment tensor $\mathbf{a}$. In addition to an ordering of the molecular axis, the dipole moments $\mathbf{e}_i$ may be aligned as well, yielding a non-zero average $\mathbf{d} = \langle \mathbf{e} \rangle$. The macroscopic polarization (or magnetization, respectively) of the resulting ferronematic state is defined as

$$\mathbf{P} = \bar{\rho} p^{\rm el} \mathbf{d}, \qquad (2.1)$$

where $\bar{\rho}$ is the number density and $p^{\rm el}$ is the strength of a dipole moment.

In the present case, where the particles' orientation is characterized by both, molecular axes and molecular dipole moments, the distribution depends on all three Euler angles $\boldsymbol{\Omega} = (\vartheta, \varphi, \alpha_{\rm dip})$. Ensemble averages of a quantity $A(\boldsymbol{\Omega})$ can then be calculated from the relation

$$\langle A \rangle = \int \rho(\boldsymbol{\Omega}) A(\boldsymbol{\Omega}) d\boldsymbol{\Omega}, \qquad (2.2)$$

where it is assumed that the distribution is normalized, i.e., $\int d\boldsymbol{\Omega}\, \rho(\boldsymbol{\Omega}) = 1$. A general expression for the angle dependence of the distribution is given by an orthogonal expansion into the (complete) set of rotational matrices $\mathcal{D}^{\ell}_{mm'}(\boldsymbol{\Omega})$ (see, e.g., [89]).

$$\rho(\boldsymbol{\Omega}) = \sum_{\ell m m'} f_{\ell m m'} \mathcal{D}^{\ell}_{mm'}(\boldsymbol{\Omega}) \qquad (2.3)$$

Here for the orientational distribution function the simple *ansatz*

$$\rho(\boldsymbol{\Omega}) = \rho_0 \left(1 + 3\mathbf{e} \cdot \mathbf{d} + \sqrt{\frac{15}{2}} \overline{\mathbf{u}\mathbf{u}} : \mathbf{a} \right), \qquad (2.4)$$

CHAPTER 2. POLAR HARD-ROD FLUIDS

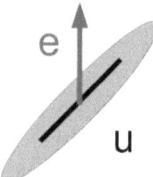

Figure 2.1: The orientation of the backbone of the molecule is related to the vector **u**. The dipole moment of the molecule is characterized by the vector **e**, that is not necessary parallel to **e**.

is employed. The second line uses the definitions of the order parameters as in (1.13) and (2.1). In the appendix the ansatz (2.4) is motivated. Note the orientational distribution function depends on two order parameters, i.e. the dipolar order parameter **d** and the quadrupolar order parameter **a** . Equation (2.4) fulfills the normalization $\int d\Omega \rho(\Omega) = 1$ since the angular integral over the resulting function $\varphi(\Omega) = 3e_\mu d_\mu + \sqrt{15/2}\,\overline{u_\mu u_\nu}a_{\mu\nu}$ vanishes. To see this explicitly, one may use that the vector components e_μ and the tensor components $u_\mu u_\nu$ are proportional to spherical harmonics Y_{lm} with $l = 1$ and $l = 2$, respectively, and $\int d\Omega Y_{lm}(\Omega) \propto \delta_{l,0}\delta_{m,0}$ [89].

2.2 Extended Potential Function for Polar Hard-Rod Fluids

The potential function for homogeneous systems is just as previously employed by [50, 51]

$$\Phi(\mathbf{a}, \mathbf{d}) = \Phi^{\mathrm{a}}(\mathbf{a}) + \Phi^{\mathrm{d}}(\mathbf{d}) + \Phi^{\mathrm{ad}}(\mathbf{a}, \mathbf{d}), \qquad (2.5)$$

where the first term corresponds to the amended potential (1.71). The second term in (2.5) is a purely polar contribution, which is modeled by

$$\Phi^{\mathrm{d}}(\mathbf{d}) = \frac{1}{2}A_{\mathrm{d}}d_\mu d_\mu - \frac{1}{4}E\ln(1 - (d_\lambda d_\lambda)^2), \qquad (2.6)$$

where A_{d} and E are parameters . In (2.6), the second term effectively limits the average dipole moment to finite values, i.e., $|\mathbf{d}|_{\max} = 1$, which is reasonable because **d** cannot increase if all dipole moments are already parallel to each other.

The specific form of $\Phi^{\mathrm{d}}(\mathbf{d})$ can be motivated as follows. Consider a system of noninteracting dipoles subject to an external electric field **E**. The corresponding (free) energy is proportional to $-\mathbf{d} \cdot \mathbf{E}$, where the magnitude of the average dipole moment $|\mathbf{d}| = \mathcal{L}(E)$, with $\mathcal{L}(x) = \coth x - 1/x$ being the Langevin function [50, 51]. An approximation of the inverse Langevin function yields $\Phi^{\mathrm{d}}(\mathbf{d})$, with $A_{\mathrm{d}} = 3$ and $E = 3$. The positive value for A_{d} implies

2.2. EXTENDED POTENTIAL FUNCTION FOR POLAR HARD-ROD FLUIDS

that the equilibrium polarization of the pure polar system is zero, corresponding to a non-ferroelectric state (note that this can change in presence of nematic ordering). For a detailed discussion see [52].

The last term on the right side of (2.5) describes the free energy contribution due to the coupling between the alignment and the polarization. To lowest order, the coupling has the form

$$\Phi^{\mathrm{ad}}(\mathbf{a}, \mathbf{d}) = c_0 \, d_\mu a_{\mu\nu} d_\nu \,. \tag{2.7}$$

A similar term was used in previous studies of polar nematics on the basis of the mesoscopic theory, where it was motivated by symmetry arguments [123, 124]. A motivation of Φ^{ad} and a relation of the coefficient c_0 to microscopic properties is based on functional arguments [125–128]. Here the derivation of [50, 51] is presented.

The orientational distribution function is used in the notation

$$\rho^{\mathrm{or}}(\boldsymbol{\Omega}) = \rho_0 \left(1 + \varphi(\boldsymbol{\Omega})\right), \tag{2.8}$$

where $\rho_0 = 1/\int d\boldsymbol{\Omega}$ corresponds to the (constant) distribution in the isotropic phase, and the function $\varphi(\boldsymbol{\Omega})$ describes the deviation from ρ_0 corresponding to anisotropic states. The free energy per particle related to the loss of orientational entropy (Δs^{or}) in anisotropic states as compared to the isotropic state is given by

$$\begin{aligned} \frac{f^{\mathrm{ent}}}{k_{\mathrm{B}} T} &= -\Delta s^{\mathrm{or}} / k_{\mathrm{B}} = \int d\boldsymbol{\Omega} \, \rho(\boldsymbol{\Omega}) \ln\left(\frac{\rho(\boldsymbol{\Omega})}{\rho_0}\right) \\ &= \rho_0 \int d\boldsymbol{\Omega} \, (1 + \varphi(\boldsymbol{\Omega})) \ln(1 + \varphi(\boldsymbol{\Omega})), \end{aligned} \tag{2.9}$$

where (2.8) is used. By definition, f^{ent} vanishes in isotropic states where $\rho(\boldsymbol{\Omega}) = \rho_0$ (and the angle-dependent deviation $\varphi(\boldsymbol{\Omega}) = 0$). For small non-zero deviations the expansion up to the third order of the integrand in (2.9) yields

$$(1 + \varphi(\boldsymbol{\Omega})) \ln(1 + \varphi(\boldsymbol{\Omega})) \approx \varphi(\boldsymbol{\Omega}) + \frac{1}{2} \varphi(\boldsymbol{\Omega})^2 - \frac{1}{6} \varphi(\boldsymbol{\Omega})^3 + \mathcal{O}(\varphi^4), \tag{2.10}$$

which gives

$$\frac{f^{\mathrm{ent}}}{k_{\mathrm{B}} T} \approx \rho_0 \left(\int d\boldsymbol{\Omega} \, \varphi(\boldsymbol{\Omega}) + \frac{1}{2} \int d\boldsymbol{\Omega} \, \varphi(\boldsymbol{\Omega})^2 - \frac{1}{6} \int d\boldsymbol{\Omega} \, \varphi(\boldsymbol{\Omega})^3 \right). \tag{2.11}$$

In each term on the right side of (2.11), the the explicit ansatz for $\varphi(\boldsymbol{\Omega})$ is inserted, i.e. $\varphi(\boldsymbol{\Omega}) = 3e_\mu d_\mu + \sqrt{15/2} \, \overline{u_\mu u_\nu} a_{\mu\nu}$. The first integral vanishes due to the normalization condition. The second integral in (2.11) may be sub-divided according to the three contributions involved in φ^2. Two of these terms are quadratic in $\mathbf{d}$ and $\mathbf{a}$, respectively, with the corresponding coefficients being non-zero. However, these terms need not to be considered further since they are not relevant for the desired coupling. These terms may be considered to be absorbed in the corresponding quadratic terms in Φ^{a} and Φ^{d}, respectively. The remaining second-order term linearly couples polarization and alignment, and thus vanishes by symmetry. Therefore, the first relevant contribution is of third order, and is given by

$$\frac{f^{\mathrm{ent}}}{k_{\mathrm{B}} T} = -\frac{9}{2} \sqrt{\frac{15}{2}} \rho_0 \, d_\mu d_\nu a_{\gamma\delta} \int d\boldsymbol{\Omega} \, e_\mu e_\nu \overline{u_\gamma u_\delta}. \tag{2.12}$$

CHAPTER 2. POLAR HARD-ROD FLUIDS

The angular integral can be calculated using the identity [89, 92] for the symmetric traceless tensors $A_{\mu\nu}$ and $B_{\gamma\delta}$,

$$\rho_0 \int d\Omega\, A_{\mu\nu} B_{\gamma\delta} = \frac{1}{5} A_{\lambda\kappa} B_{\lambda\kappa} \triangle_{\mu\nu,\gamma\delta} \,, \tag{2.13}$$

where

$$\triangle_{\mu\nu,\gamma\delta} = \frac{1}{2}\left(\delta_{\mu\gamma}\delta_{\nu\delta} + \delta_{\mu\delta}\delta_{\nu\gamma}\right) - \frac{1}{3}\delta_{\mu\nu}\delta_{\gamma\delta} \tag{2.14}$$

is the isotropic 4-th rank tensor with the appropriate symmetry. Notice that $\triangle_{\lambda\kappa,\lambda\kappa} = 5$. Applying (2.13) in (2.12) one finally obtains

$$\frac{f^{\text{ent}}}{k_{\text{B}}T} = \frac{c_0}{2} d_\mu a_{\mu\nu} d_\nu = \Phi^{\text{ad}}(\mathbf{a},\mathbf{d}), \tag{2.15}$$

where the prefactor

$$c_0 = -3\sqrt{\frac{6}{5}} P_2(\mathbf{e}\cdot\mathbf{u}) = -3\sqrt{\frac{6}{5}} P_2(\cos\alpha_{\text{dip}}) \tag{2.16}$$

depends on the angle between the molecular axis and the molecular dipole moment. For a more detailed description it is refereed to [52].

In systems where the particles are characterized by longitudinal (or nearly longitudinal) dipoles (i.e., $\alpha_{\text{dip}} \approx 0$ or $\alpha_{\text{dip}} \approx \pi$) the coefficient c_0 becomes negative, implying that the potential favors macroscopic polarization *parallel* (or antiparallel) to the director. The opposite situation occurs for transversal dipoles (i.e., $\alpha_{\text{dip}} \approx \pi/2$) where the free energy favors perpendicular orientation of $\mathbf{d}$ and the director. The present work is focused on the case of longitudinal dipoles.

The potential function (2.5) is appropriate to model the equilibrium behavior of homogeneous polar rod dispersions. For heterogeneous systems additional terms are necessary.

The elastic contribution to the free energy is modeled by the Frank elasticity in the one constant approximation (1.41). Within the Landau-Ginzburg theory similar gradient terms occur for the average dipole moment $\mathbf{d}$. The lowest order term is $(\nabla_\mu d_\nu)(\nabla_\mu d_\nu)$. In addition a coupling between gradients of the alignment tensor $\nabla\mathbf{a}$ and the polarization vector $\mathbf{d}$ is possible. The induced polarization by orientational distorsions referred to as flexoelectric effect is investigated theoretically and measured in experiments. [7, 129–133]. The inverse effect where the alignment tensor field is distorted by external electric fields is also observable. For simplicity it is considered an approximation similar to the one constant approximation, where the anisotropy of this effect is disregard. The flexoelectric effect is modeled by the term $c_f d_\mu \nabla_\nu a_{\mu\nu}$. To sum up, the total potential reads

$$\Phi^{\text{tot}}(\mathbf{d},\mathbf{a}) = \Phi^{\text{a}} + \frac{1}{2}\xi_{\text{a}}^2 \nabla_\mu a_{\nu\rho} \nabla_\mu a_{\nu\rho} + \Phi^{\text{d}} + \frac{1}{2}\xi_{\text{d}}^2 \nabla_\mu d_\nu \nabla_\mu d_\nu + \Phi^{\text{ad}} - c_{\text{f}} d_\mu \nabla_\nu a_{\mu\nu}. \tag{2.17}$$

Note, the ordinary flexoelectric effect is related to external fields. Here it is assumed that the internal average dipole moment $\mathbf{d}$ acts in a similar way on the alignment.

2.3 Relaxation Equation and Constitutive Pressure Tensor Equation for Polar Hard-Rod Fluids

The entropy production is used as a guideline for setting up the constitutive equations for the friction pressure tensor and for the relaxation equation for the alignment tensor it is assumed as in [45], that the generalized Gibbs relation

$$\frac{ds}{dt} = T^{-1}\left(\frac{du}{dt} + p\frac{d\rho^{-1}}{dt}\right) - T^{-1}\frac{dg}{dt} \tag{2.18}$$

holds true for dynamic phenomena. The specific Gibbs free potential $g(\mathbf{a}, \nabla \mathbf{a}, \mathbf{d}, \nabla \mathbf{d})$ is associated with alignment and the averaged dipole moment. Here the simple ansatz is employed

$$g(\mathbf{a}, \nabla \mathbf{a}, \mathbf{d}, \nabla \mathbf{d}) = \frac{k_B T}{m}\Phi^{\text{tot}}. \tag{2.19}$$

To obtain the relaxation equation for the alignment tensor as well as for the dipole vector the balance equations for $\mathbf{a}$ and $\mathbf{d}$ are needed. The balance equation for the alignment tensor reads [113]

$$\frac{d}{dt}a_{\mu\nu} - 2\overline{\varepsilon_{\mu\lambda\rho}\omega_\lambda a_{\rho\nu}} - 2\kappa_a\overline{\Gamma_{\mu\lambda}a_{\lambda\nu}} + \nabla_\lambda b_{\lambda\mu\nu} = \left(\frac{\delta a_{\mu\nu}}{\delta t}\right)_{irr}, \tag{2.20}$$

where $b_{\lambda\mu\nu}$ is the alignment flux tensor. Similarly, the balance equation of the dipole vector is chosen as

$$\frac{d}{dt}d_\mu - \varepsilon_{\mu\lambda\rho}\omega_\lambda d_\rho - \kappa_d \Gamma_{\mu\lambda}d_\lambda + \nabla_\lambda c_{\lambda\mu} = \left(\frac{\delta d_\mu}{\delta t}\right)_{irr}. \tag{2.21}$$

The entropy production is given by

$$\frac{ds}{dt} = T^{-1}\left(\frac{du}{dt} + p\frac{d\rho^{-1}}{dt}\right) - \frac{k_B}{m}g_{\mu\nu}\frac{d}{dt}a_{\mu\nu} - \frac{k_B}{m}g_\mu\frac{d}{dt}d_\mu.$$

The tensor $g_{\mu\nu}$ and the vector g_μ refers to the derivative of the specific Gibbs free energy of the alignment tensor and the the dipole moment, respectively

$$g_{\mu\nu} = \frac{\delta g}{\delta a_{\mu\nu}} = \Phi_{\mu\nu}^{a} - \xi_a^2 \triangle a_{\mu\nu} + c_f\overline{\nabla_\mu d_\nu} + \frac{1}{2}c_0\overline{d_\mu d_\nu} \tag{2.22}$$

$$g_\mu = \frac{\delta g}{\delta d_\mu} = \Phi_\mu^{d} - \xi_d^2 \triangle d_\mu - c_f \nabla_\lambda a_{\lambda\mu} + c_0 a_{\mu\lambda}d_\lambda.$$

CHAPTER 2. POLAR HARD-ROD FLUIDS

The entropy production related to the anisotropic contribution is given by

$$\begin{aligned}\rho\left(\frac{\delta s}{\delta t}\right)_{\text{aniso}} &= -\frac{k_B T}{m}\left[g_{\mu\nu}\left(\left(\frac{\delta a_{\mu\nu}}{\delta t}\right)_{\text{irr}} + 2\,\overline{(\varepsilon_{\mu\sigma\lambda}\omega_\sigma a_{\lambda\mu})} \right) + 2\kappa_a\,\overline{\Gamma_{\mu\lambda}a_{\lambda\nu}} - \nabla_\lambda b_{\lambda\mu\nu}\right) \\ &\quad + g_\mu\left(\left(\frac{\delta d_\mu}{\delta t}\right)_{\text{irr}} + \varepsilon_{\mu\lambda\sigma}\omega_\lambda d_\sigma + 2\kappa_d\,\Gamma_{\mu\lambda}d_\lambda - \nabla_\lambda c_{\lambda\mu}\right)\right] \\ &= -\frac{k_B T}{m}\left[g_{\mu\nu}\left(\left(\frac{\delta a_{\mu\nu}}{\delta t}\right)_{\text{irr}} + 2\,\overline{(\varepsilon_{\mu\sigma\lambda}\omega_\sigma a_{\lambda\mu})}\right) + 2\kappa_a\,\overline{\Gamma_{\mu\lambda}a_{\lambda\nu}}\right) \\ &\quad - \nabla_\lambda g_{\mu\nu}b_{\lambda\mu\nu} + b_{\lambda\mu\nu}\nabla_\lambda g_{\mu\nu} \\ &\quad + g_\mu\left(\left(\frac{\delta d_\mu}{\delta t}\right)_{\text{irr}} + \varepsilon_{\mu\lambda\sigma}\omega_\lambda d_\sigma + 2\kappa_d\,\Gamma_{\mu\lambda}d_\lambda\right) - \nabla_\lambda g_\mu c_{\lambda\mu} + c_{\lambda\mu}\nabla_\lambda g_\mu\right], \end{aligned} \quad (2.23)$$

where the divergence terms $\nabla_\lambda g_\mu c_{\lambda\mu}$ and $\nabla_\lambda g_{\mu\nu}b_{\lambda\mu\nu}$ are related to the entropy flux $s_\lambda = ...g_\mu c_{\lambda\mu} + ...g_{\mu\nu}b_{\lambda\mu\nu}$. Since the terms involving ω are reversible the anisotropic irreversible contribution of second rank tensors to the entropy production yields

$$\left(\rho\frac{\delta s}{\delta t}\right)^{(2)}_{\text{irr(aniso)}} = g_{\mu\nu}\left(\frac{\delta a_{\mu\nu}}{\delta t}\right)_{\text{irr}} + 2\kappa_a\,g_{\mu\nu}\overline{\Gamma_{\mu\lambda}a_{\lambda\nu}} + c_{\lambda\mu}\nabla_\lambda g_\mu + 2\kappa_d g_\mu d_\nu \Gamma_{\mu\nu}. \quad (2.24)$$

The total irreversible contribution to the entropy production involving second rank tensors is given by

$$\begin{aligned}\rho T\left(\frac{\delta s}{\delta t}\right)^{(2)}_{\text{irr}} &= -\overline{p_{\mu\nu}\nabla_\mu v_\nu} \\ &\quad - \frac{\rho}{m}k_B T\left(g_{\mu\nu}\left[\left(\frac{\delta a_{\mu\nu}}{\delta t}\right)_{\text{irr}} + 2\kappa_a\overline{\Gamma_{\mu\lambda}a_{\lambda\nu}}\right] + c_{\lambda\mu}\nabla_\lambda g_\mu + 2\kappa_d g_\mu d_\nu \Gamma_{\mu\nu}\right) \\ &= -\overline{p_{\mu\nu}\nabla_\mu v_\nu} - \frac{\rho}{m}k_B T\left(2\kappa_a\overline{a_{\mu\lambda}g_{\lambda\nu}} + 2\kappa_d g_\mu d_\nu\right)\overline{\nabla_\mu v_\nu} \\ &\quad - \frac{\rho}{m}k_B T\left(g_{\mu\nu}\left(\frac{\delta a_{\mu\nu}}{\delta t}\right)_{\text{irr}} + c_{\lambda\mu}\nabla_\lambda g_\mu\right), \end{aligned} \quad (2.25)$$

where first term in the first line is the isotropic contribution [47] and the balance equation for the alignment tensor (2.20) was used. Under the assumption that the system is not too far from equilibrium tensorial forces are linear functions of tensorial fluxes [112] and the following equations result:

$$-g_{\mu\nu} = \tau_a\left(\frac{\delta a_{\mu\nu}}{\delta t}\right)_{\text{irr}} \quad (2.26)$$
$$+ \sqrt{2}\tau_{ap}\overline{\nabla_\mu v_\nu} + \ell_{ac}\overline{\nabla_\mu g_\nu}$$

$$-\overline{p}_{\mu\nu} - p_{\text{kin}}\left(2\kappa_a\overline{a_{\mu\lambda}g_{\lambda\nu}} + 2\kappa_d\overline{g_\mu d_\nu}\right) = \sqrt{2}\tau_{pa}p_{\text{kin}}\left(\frac{\delta a_{\mu\nu}}{\delta t}\right)_{\text{irr}} \quad (2.27)$$
$$+ 2\tau_p p_{\text{kin}}\overline{\nabla_\mu v_\nu} + \ell_{pc}p_{\text{kin}}\overline{\nabla_\mu g_\nu} \quad (2.28)$$

$$-c_{\mu\nu} = \ell_{ca}\left(\frac{\delta a_{\mu\nu}}{\delta t}\right)_{\text{irr}} + \sqrt{2}\ell_{cp}\overline{\nabla_\mu v_\nu} + \tilde{\mu}\overline{\nabla_\mu g_\nu},$$

2.3. RELAXATION EQUATION AND CONSTITUTIVE PRESSURE TENSOR EQUATION FOR POLAR HARD-ROD FLUIDS

with the Onsager symmetry relations [112] $\tau_{ap} = \tau_{pa}$, $\ell_{ac} = \ell_{ca}$, $\ell_{pc} = \ell_{cp}$, positive entropy production imposes the inequalities $\tau_a \geq 0$, $\tau_p \geq 0$, $\tilde{\mu} \geq 0$ and the following relations $\tau_a \tau_p \geq (\tau_{ap})^2$, $\tau_a \tilde{\mu} \geq (\ell_{ac})^2$ and $\tau_p \tilde{\mu} \geq (\ell_{pc})^2$.

Insertion of the quantity $(\delta a_{\mu\nu}/\delta t)_{irr}$, as inferred from (2.20), into the Eq.(2.26) yields the relaxation equation

$$\frac{d}{dt}a_{\mu\nu} - 2\overline{\epsilon_{\mu\lambda\rho}\omega_\lambda a_{\rho\nu}} - 2\kappa_a \overline{\Gamma_{\mu\lambda}a_{\lambda\nu}} + \nabla_\lambda b_{\lambda\mu\nu} = \qquad (2.29)$$
$$-\tau_a^{-1}g_{\mu\nu} - \sqrt{2}\frac{\tau_{ap}}{\tau_a}\Gamma_{\mu\nu} - \frac{\tau_{ac}}{\tau_a}\overline{\nabla_\mu g_\nu}.$$

To derive the relaxation equation for the average dipole moment the irreversible contribution to the entropy production involving vectors is considered, i.e.

$$\left(\frac{\delta}{\delta t}d_\mu\right)_{irr} = -\tau_d^{-1}g_\mu. \qquad (2.30)$$

Applying the balance equation for the average dipole moment (2.21) in (2.30) one obtains the relaxation equation

$$\frac{d}{dt}d_\mu - \epsilon_{\mu\lambda\rho}\omega_\lambda d_\rho - 2\kappa_d \Gamma_{\mu\lambda}d_\lambda + \nabla_\lambda c_{\lambda\mu} = -\tau_d^{-1}g_\mu. \qquad (2.31)$$

The use of Eq. (2.26) and Eq. (2.29) leads to the expression

$$c_{\mu\nu} = \hat{\mu}\overline{\nabla_\mu g_\nu} + \sqrt{2}\ell\overline{\nabla_\mu v_\nu} - \frac{\ell_{ca}}{\tau_a}g_{\mu\nu}, \qquad (2.32)$$

with abbreviation

$$\hat{\mu} = \left(\tilde{\mu} - \frac{\ell_{ca}\ell_{ac}}{\tau_a}\right) \quad \text{and} \quad \ell = \left(\sqrt{2}\ell_{cp} - \frac{\ell_{ca}\tau_{ap}}{\tau_a}\right). \qquad (2.33)$$

The constitutive equation for the pressure tensor with (2.29) and (2.27) yields

$$\overline{p}_{\mu\nu} = -2\eta_{iso}\overline{\nabla_\mu v_\nu} + \frac{\rho}{m}k_BT\left(\sqrt{2}\frac{\tau_{pa}}{\tau_a}g_{\mu\nu} - 2\kappa_a\overline{a_{\mu\lambda}g_{\lambda\nu}} - 2\kappa_d\overline{g_\mu d_\nu}\right) - \eta_d\overline{\nabla_\mu g_\nu}, \qquad (2.34)$$

where

$$\eta_{iso} = p_{kin}\left(\tau_p - \frac{\tau_{ap}\tau_{pa}}{\tau_a}\right) \quad \text{and} \quad \eta_d = \sqrt{2}p_{kin}\ell. \qquad (2.35)$$

The first term of (2.34) already occurs for simple fluids the second and third term existent for fluids with an orientational degree of freedom and the last two terms results from the additional polar coupling.

2.4 Scaled Variables

The model equations for polar rod dispersions are scaled according to the shear rate scaling cf. (1.5.2). The alignment tensor and the average dipole vector are expressed in units of the value of the nematic order parameter a_K at the isotropic-nematic phase transition temperature, T_K, yielding $\mathbf{a}^* = \mathbf{a}/a_K$ and $\mathbf{d}^* = \mathbf{d}/a_K$.

To treat the polar contribution to the potential [see (2.6)], in analogy to the alignment case, the scaled "dipole temperature" $\vartheta_d = a_K \Phi_{\text{ref}}^{-1} A_d = \delta_K^{-1} A_0^{-1} A_d$ is introduced. The dimensionless parameter $E^* = a_K^3 \Phi_{\text{ref}}^{-1} E = a_K^2 \delta_K^{-1} A_0^{-1} E$. The polar contribution then becomes

$$\Phi^{d*}(\mathbf{d}^*) = \frac{1}{2} \vartheta_d d_\mu^* d_\mu^* - \frac{1}{4} E^* \ln(1 - (d_\lambda^* d_\lambda^*)^2). \tag{2.36}$$

The coupling term (2.7) is made dimensionless by introducing the parameter $c = c_0 a_K^2 \Phi_{\text{ref}}^{-1}$, yielding

$$\Phi^{ad*}(\mathbf{a}^*, \mathbf{d}^*) = \frac{c}{2} d_\mu^* a_{\mu\nu}^* d_\nu^*. \tag{2.37}$$

The relaxation equation for the alignment tensor in scaled form reads

$$\frac{d}{dt^*} a_{\mu\nu}^* = 2\overline{\varepsilon_{\mu\lambda\rho} \omega_\lambda^* a_{\rho\nu}^*} + 2\kappa_a \overline{\Gamma_{\mu\lambda}^* a_{\lambda\nu}^*} + \bar{D} \triangle^* \Phi_{\mu\nu}^{*a} + \frac{1}{Er} \triangle^* a_{\mu\nu}^* \tag{2.38}$$
$$- \frac{\bar{D} Wi}{Er} \triangle^{2*} a_{\mu\nu}^* - \frac{1}{Wi} \Phi_{\mu\nu}^{a*} + \sqrt{\frac{3}{2}} \lambda_K \Gamma_{\mu\nu}^* - \frac{c}{Wi} \overline{d_\mu^* d_\nu^*} - \mathcal{E}_{fa} \overline{\nabla_\mu^* d_\nu^*}$$
$$- \lambda_d \left(\frac{1}{Wi_d} \overline{\nabla_\mu^* \Phi_\nu^{d*}} - \frac{1}{Er_d} \triangle^* \overline{\nabla_\mu^* d_\nu^*} - \mathcal{E}_{fd} \overline{\nabla_\mu^* \nabla_\lambda^* a_{\lambda\nu}^*} + \frac{c}{Wi_d} \overline{\nabla_\mu^* a_{\nu\lambda}^* d_\lambda^*} \right).$$

Similarly, the relaxation equation for the averaged dipole moment is given by

$$\frac{d}{dt^*} d_\mu^* = 2\varepsilon_{\mu\lambda\rho} \omega_\lambda^* d_\rho^* + 2\kappa_a \Gamma_{\mu\lambda}^* d_\lambda^* - \frac{1}{Wi_d} \Phi_\mu^{d*} + \frac{1}{Er_d} \triangle^* d_\mu^* \tag{2.39}$$
$$- \frac{c}{Wi_d} a_{\mu\lambda}^* d_\lambda^* + \mathcal{E}_{fd} \nabla_\lambda^* a_{\lambda\mu}^* - \zeta_1 \nabla_\lambda^* g_{\lambda\mu}^* - \zeta_2 \triangle^* v_\mu^* + \zeta_3 \triangle^* g_\mu^*.$$

The corresponding parameters are the Weissenberg number related to $\mathbf{a}$ and to $\mathbf{d}$, respectively

$$Wi = \frac{\tau_a}{A_K} \dot{\gamma}^{\text{eff}} = \frac{1}{6\bar{D}_R A_K} \dot{\gamma}^{\text{eff}}, \quad Wi_d = \frac{\tau_d}{A_K} \dot{\gamma}^{\text{eff}}, \tag{2.40}$$

the Ericksen number related to $\mathbf{a}$ and to $\mathbf{d}$, respectively

$$Er = Wi \left(\frac{2h}{\xi_a} \right)^2 A_K, \quad Er_d = Wi_d \left(\frac{2h}{\xi_d} \right)^2 A_K, \tag{2.41}$$

the dimensionless flexoelectric parameters

$$\mathcal{E}_{fa} = \frac{c_f}{\tau_a u^w p^{el} |\mathbf{d}|}, \quad \mathcal{E}_{fd} = \frac{c_f}{\tau_d u^w p^{el} |\mathbf{d}|}, \tag{2.42}$$

2.4. SCALED VARIABLES

Parameters	a	d
Weissenberg number	$Wi = \frac{\tau_a}{A_K}\dot{\gamma}^{\text{eff}}$	$Wi_d = \frac{\tau_a}{A_K}\dot{\gamma}^{\text{eff}}$
Ericksen number	$Er = \tau_a \dot{\gamma}^{\text{eff}} \left(\frac{2h}{\xi_a}\right)^2$	$Er_d = \tau_d \dot{\gamma}^{\text{eff}} \left(\frac{2h}{\xi_d}\right)^2$
λ_K		$-\frac{2}{3}\frac{\tau_{ap}}{\tau_a a_K}$
ν_{iso}		$\frac{\eta_{iso}}{p_{kin}a_K^2 A_K}\dot{\gamma}^{eff}$
β		$\frac{\rho(v^w)^2}{p_{kin}a_K^2 A_K}$

Table 2.1: The parameters related to the alignment tensor **a** and the average dipole moment **d** are displayed.

the tumbling parameter

$$\lambda_K = -\frac{2}{\sqrt{3}}\frac{\tau_{ap}}{\tau_a a_K} \tag{2.43}$$

and the parameters

$$\lambda_d = \frac{\ell_{ac}A_K}{u^w \tau_a}, \quad \zeta_1 = \frac{\ell_{ca}A_K}{\tau_a u^w}, \quad \zeta_2 = \frac{\sqrt{2}\ell}{2h\, a_K}, \quad \zeta_3 = \frac{A_K \bar{\mu}}{2h\, u^w}. \tag{2.44}$$

Similarly, the constitutive equation for the pressure tensor is scaled, viz.

$$\overline{p}^*_{\mu\nu} = -2\nu_{\text{iso}}\overline{\nabla^*_\mu v^*_\nu} - \nu_d \overline{\nabla^*_\mu g^*_\nu} - \sqrt{\frac{3}{2}}\lambda_K g^*_{\mu\nu} - 2\kappa_a \overline{a^*_{\mu\lambda} g^*_{\lambda\nu}} - 2\bar{\kappa}_d \overline{g^*_\mu d^*_\nu}, \tag{2.45}$$

with

$$g^*_{\mu\nu} = \frac{1}{Wi}\Phi^{a*}_{\mu\nu} - \frac{1}{Er}\triangle^* a^*_{\mu\nu} + \mathcal{E}_{\text{fa}}\overline{\nabla^*_\mu d^*_\nu} + \frac{c}{Wi}\overline{d^*_\mu d^*_\nu} \tag{2.46}$$

$$g^*_\mu = \frac{1}{Wi_d}\Phi^{d*}_\mu - \frac{1}{Er_d}\triangle^* d^*_\mu - \mathcal{E}_{\text{fd}}\nabla^*_\lambda a^*_{\lambda\mu} + \frac{c}{Wi_d}a^*_{\mu\lambda}d^*_\lambda \tag{2.47}$$

and the scaled parameters

$$\nu_{\text{iso}} = \frac{\eta_{\text{iso}} v^w}{p_{\text{kin}} a_K^2 A_K}, \quad \nu_d = \frac{\eta_d}{a_K^2 A_K p_{\text{kin}} 2h}, \quad \bar{\kappa}_d = \frac{\kappa_d}{a_K}A_K. \tag{2.48}$$

For simplicity the asterisks for the scaled variables are omitted in the following. The analysis of polar rod dispersions under shear is simplified by disregarding the alignment tensor flux and average dipole moment flux contributions, i.e. the parameters ζ_i and $\bar{D}$ are set to be zero. In the table (2.1) the parameter and scaled variables are summarized.

CHAPTER 2. POLAR HARD-ROD FLUIDS

2.5 Component Form of the Model Equations

As in section (1.7) the model equations can be expressed in the tensor basis (1.32). From Eq. (2.38) one obtains a system of nine partial differential equations. The component form of the alignment tensor is given by

$$
\begin{aligned}
\dot{a}_0 &= -\frac{1}{Wi}\hat{\Phi}_0^a - \frac{1}{3}\sqrt{3}\,\kappa_a\,a_2\,\partial_y u, +\mathcal{E}_{\text{fa}}\frac{1}{\sqrt{6}}\partial_y d_2 - \frac{1}{Er}\partial_y^2 a_0 \\
\dot{a}_1 &= -\frac{1}{Wi}\hat{\Phi}_1^a + \partial_y u\,a_2 + \mathcal{E}_{\text{fa}}\frac{1}{\sqrt{2}}\partial_y d_2 - \frac{1}{Er}\partial_y^2 a_1, \\
\dot{a}_2 &= -\frac{1}{Wi}\hat{\Phi}_2^a - \partial_y u\,a_1 + \frac{\sqrt{3}}{2}\lambda_{\text{K}}\,\partial_y u - \frac{1}{3}\sqrt{3}\,\kappa_a\,a_0\,\partial_y u \\
&\quad - \mathcal{E}_{\text{fa}}\frac{1}{\sqrt{2}}\partial_y d_1 - \frac{1}{Er}\partial_y^2 a_2, \qquad (2.49)\\
\dot{a}_3 &= -\frac{1}{Wi}\hat{\Phi}_3^a + \frac{1}{2}a_4\,(\kappa_a + 1)\,\partial_y u - \frac{1}{Er}\partial_y^2 a_3, \\
\dot{a}_4 &= -\frac{1}{Wi}\hat{\Phi}_4^a + \frac{1}{2}a_3(\kappa_a - 1)\,\partial_y u - \mathcal{E}_{\text{fa}}\frac{1}{\sqrt{2}}\partial_y d_3 - \frac{1}{Er}\partial_y^2 a_4,
\end{aligned}
$$

where the potential is given by

$$
\begin{aligned}
\hat{\Phi}_0^a &= \Phi_0^a - \frac{c}{2\sqrt{6}}\left(d_1^2 + d_2^2 - 2d_z^2\right), \\
\hat{\Phi}_1^a &= \Phi_1^a + \frac{c}{2\sqrt{2}}\left(d_1^2 - d_2^2\right), \\
\hat{\Phi}_2^a &= \Phi_2^a + \frac{c}{\sqrt{2}}d_1 d_2 \qquad (2.50)\\
\hat{\Phi}_3^a &= \Phi_3^a + \frac{c}{\sqrt{2}}d_1 d_3 \\
\hat{\Phi}_4^a &= \Phi_4^a + \frac{c}{\sqrt{2}}d_2 d_3.
\end{aligned}
$$

The component form of the potential Φ^a is given in (1.85). The partially differential equations for the averaged dipole moment components in scaled form (2.39) reads

$$
\begin{aligned}
\dot{d}_1 &= -\frac{1}{3}\hat{\Phi}_1^d + \kappa_d\,d_2\,\partial_y u - \mathcal{E}_{\text{fd}}\frac{1}{\sqrt{2}}\partial_y a_2, +\frac{1}{Er_d}\partial_y^2 d_1 \\
\dot{d}_2 &= -\frac{1}{3}\hat{\Phi}_2^d + \kappa_d\,d_1\,\partial_y u + \mathcal{E}_{fd}\partial_y(\frac{1}{\sqrt{6}}a_0 + \frac{1}{\sqrt{2}}a_1) + \frac{1}{Er_d}\partial_y^2 d_2 \qquad (2.51)\\
\dot{d}_3 &= -\frac{1}{3}\hat{\Phi}_3^d - \mathcal{E}_{\text{fd}}\frac{1}{\sqrt{2}}\partial_y a_4 + \frac{1}{Er_d}\partial_y^2 d_2,
\end{aligned}
$$

with the polar potential

$$\hat{\Phi}_1^d = (\vartheta_d + E^* d_k^2)d_1 + \frac{c}{\sqrt{2}}\left(-\frac{1}{\sqrt{3}}a_0 d_1 + a_1 d_1 + a_2 d_2 + a_3 d_3\right)$$

$$\hat{\Phi}_2^d = (\vartheta_d + E^* d_k^2)d_2 + \frac{c}{\sqrt{2}}\left(-\frac{1}{\sqrt{3}}a_0 d_2 - a_1 d_2 + a_2 d_1 + a_4 d_3\right) \quad (2.52)$$

$$\hat{\Phi}_3^d = (\vartheta_d + E^* d_k^2)d_3 + \frac{c}{\sqrt{2}}\left(\frac{2}{\sqrt{3}}a_0 d_3 + a_3 d_1 + a_4 d_2\right)$$

and the abbreviation $d_k^2 = d^2/(1 - d^4/d_{\max}^4)$. Here $d_{\max}$ is assumed to be one. The notation $d^2 \equiv d_1^2 + d_2^2 + d_3^2$ was used. The momentum equation in the component form reads

$$\begin{aligned}
\frac{\partial u}{\partial t} &= \frac{\nu_{\mathrm{iso}}}{\beta}\partial_y^2 u + \sqrt{\frac{3}{2}}\frac{\lambda_K}{\beta}\partial_y \Phi_2^a - \sqrt{\frac{3}{2}}\frac{\lambda_K}{\beta}\frac{Wi}{Er}\partial_y^3 a_2 \\
&+ \frac{\sqrt{3}}{2}\frac{\lambda_K}{\beta}c\,\partial_y(d_1 d_2) + \frac{\sqrt{3}}{2}\frac{\lambda_K}{\beta}\mathcal{E}_{\mathrm{fa}}\partial_y^2 d_1 \\
&+ \frac{\kappa}{\beta}\partial_y\left(\frac{1}{2\sqrt{2}}a_3\,\Phi_4^a - \frac{1}{2\sqrt{2}}\frac{Wi}{Er}a_3\,\partial_y^2 a_4 - \frac{1}{\sqrt{6}}a_0\Phi_2^a + \frac{1}{\sqrt{6}}\frac{Wi}{Er}a_0\,\partial_y^2 a_2 \right. \\
&\left. \quad - \frac{c}{2\sqrt{3}}a_0\partial_y(d_1 d_2) - \frac{c}{8\sqrt{3}}\mathcal{E}_{\mathrm{fa}}\,a_0\,\partial_y^2 d_1 \partial_y(d_2 d_3) + \frac{1}{4}\mathcal{E}_{\mathrm{fa}}a_3\,\partial_y^2 d_3\right) \\
&- \frac{1}{2}\frac{\eta_d}{Wi_d}\partial_y \Phi_1^d + \frac{\eta_d}{2Er_d}\partial_y^3 d_1 - \frac{1}{2\sqrt{2}}\eta_d\mathcal{E}_{\mathrm{fd}}\partial_y^2 a_2 \\
&- \frac{c\eta_d}{Wi_d}\partial_y\left(\frac{1}{\sqrt{2}}a_2 d_1 - \frac{1}{\sqrt{2}}a_1 d_2 - \frac{1}{\sqrt{6}}a_0\,d_2 + \frac{1}{\sqrt{2}}a_4\,d_3\right).
\end{aligned}$$

2.6 Magnetic Fields

A time-dependent average dipole moment in the rod system generates electrodynamic fields. Systems with polarization $\mathbf{P}(t)$ give rise to a magnetic field $\mathbf{H}(t) = \mu_0^{-1}\mathbf{B}(t)$. Note, however, that analogous considerations can be made for systems with magnetic molecular dipoles leading to an electric field.

According to Maxwell's relations the magnetic field fulfills, in the absence of currents,

$$\nabla \times \mathbf{H} = \frac{\partial}{\partial t}\mathbf{D}, \quad (2.54)$$

where $\mathbf{D} = \epsilon_0 \mathbf{E} + \mathbf{P}$ is the dielectric displacement. In the following the (possibly non-zero) electric field corresponding to the internal field in a ferronematic state is neglected, i.e. set $\mathbf{E} = \mathbf{0}$, yielding $\partial \mathbf{D}/\partial t = \partial \mathbf{P}/\partial t = \dot{\mathbf{P}}$.

To estimate the magnetic field the Maxwell equation (2.54) is integrated over a surface A with boundary ∂A, yielding

$$\int_{\partial A} \mathbf{H}\cdot d\mathbf{s} = \frac{d}{dt}\int_A \mathbf{D}\cdot d\mathbf{f} = \int_A \dot{\mathbf{P}}\cdot d\mathbf{f}, \quad (2.55)$$

CHAPTER 2. POLAR HARD-ROD FLUIDS

where the Stokes' law is used on the left side, and **s** and **f** are the tangential and surface normal vector, respectively. A rectangular surface located in the shear (x-y) plane is chosen for the surface A. The upper and lower limits of y are given by the positions of the two plane plates of the Couette shear cell, $y_{\min} = -h$ and $y_{\max} = h$, and the length L in x direction is arbitrary.

It is assumed that the plates are large compared to the separation $2h$. The magnetic field **H** only depends on y, because translation in the x-z-plane parallel to the plates provides no changes in the geometry. Equation (2.55) then simplifies to

$$\int_0^L dx \; (H_x(x, y_{\min}) - H_x(x, y_{\max})) \qquad (2.56)$$
$$+ \int_{y_{\min}}^{y_{\max}} dy \; (H_y(x_{\min}, y) - H_y(x_{\max}, y)) = L \int_{y_{\min}}^{y_{\max}} \dot{\mathbf{P}} \cdot \hat{\mathbf{z}} \, dy \, .$$

The second integral vanishes since the magnetic field does not depend on the variable x. Thus, (2.56) reduces to

$$H_x(h, t) - H_x(-h, t) = -\int_{-h}^{h} \dot{P}_z(y, t) dy \, , \qquad (2.57)$$

where H_α is the α-component of **H**, and the fact that the length L cancels on both sides was used. An analogous equation for $H_z(h)$ is obtained by an integration over the y-z plane.

Further relations for the magnetic field can be derived on account of the fact that the Maxwell equation(2.54), which involves one spatial derivative, is parity invariant. Thus, the magnetic field itself must have odd parity, that is $\mathbf{H}(y) = -\mathbf{H}(-y)$. Therefore, (2.57) implies the relation $2H_x(h) = -\int_{-h}^{h} \dot{P}_z(y) dy$, and an analogous relation holds for $H_z(h)$. Taken altogether, the magnetic flux $\mathbf{B} = \mu_0 \mathbf{H}$ immediately above the upper plate is given by

$$B_x(t) = -\frac{\mu_0}{2} \int_{-h}^{h} \dot{P}_z(y, t) \, dy, \quad B_z(t) = \frac{\mu_0}{2} \int_{-h}^{h} \dot{P}_x(y, t) \, dy \, . \qquad (2.58)$$

Part III

Applications

3

Orientational Bulk Dynamics of Non-Polar Hard-Rod Fluids

In this chapter the monodomain response of nonpolar rods subjected to a steady flow is presented. Firstly, extensional flows are considered and the dependence of the order parameter on the extension rate is investigated. It reveals that the orientational response on the extension rate in not bounded when the model is based on the Landau-de Gennes potential. In contrast, solutions involving the amended potential restrict the order parameter and agree with experimental values. In the second section characteristic solutions of nonpolar rods subjected to a steady shear flow are reviewed. The last part of the chapter deals with the robustness of the characteristic solutions against shear rate fluctuations.

3.0.1 Extensional Flows

In section (1.6) the amended Landau-de Gennes potential was introduced and a theoretical motivation was given. As discussed, terms of higher than 4th order are disregarded in the Landau-de Gennes potential even though these terms are important in order to restrict the range of values of the order parameter. For extensional flows in particular, this fact becomes significant. The authors of [134] analyzed the extensional flow of a two-dimensional nematic polymer. From the numerical solution $\rho^{or}(\mathbf{u})$ of the corresponding Fokker-Planck equation, these authors calculated the value of the order parameter as a function of the extension rate. Within this approach, the values of the order parameter are calculated from $\rho^{or}(\mathbf{u})$ via Eq. (1.21) and naturally obey the physical bounds.

The moment equation for the alignment tensor including the Landau-de Gennes potential, however, does not restrict the values of the orientational order parameter for extensional flows. Numerical solutions show that the dynamical equation with the amended potential proposed in Sec. 1.6 restricts the order parameter in a satisfying form.

For planar biaxial extensional flows in the x-direction and an extension rate $\dot{\varepsilon}$, the strain tensor can be written as $\Gamma_{\mu\nu} = \text{diag}(\dot{\varepsilon}, -\dot{\varepsilon}, 0)$ and therefore the dynamics of the alignment tensor components in the tensor basis (1.32) are given by

CHAPTER 3. ORIENTATIONAL BULK DYNAMICS OF NON-POLAR HARD-ROD FLUIDS

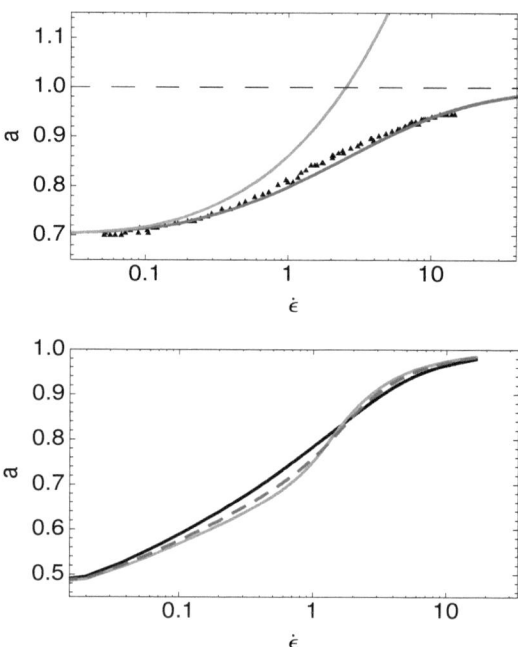

Figure 3.1: Upper graph: The average of the scalar order parameter $a = \sqrt{\mathbf{a} : \mathbf{a}}$ calculated numerically is plotted versus the extensional rate $\dot{\epsilon}/6$. The points are the results from the numerical solution of the Fokker-Planck equation obtained in [134]. The orientation increases without limits for the Landau-de Gennes potential (gray line), whereas it saturates for the amended potential (black line). The model parameter are $\lambda_K = 1$, $\kappa = 0$, and $\vartheta = 0$. Lower graph: The average of the scalar order parameter $a = \sqrt{\mathbf{a} : \mathbf{a}}$ as a function of the extensional rate is displayed for different parameter values κ_a ($\kappa_a = 0.4$ dark curve, $\kappa_a = 0.8$ dashed curve, $\kappa_a = 1$ gray curve). The model parameters are $\lambda_K = 4/3$ and $\vartheta = 0$.

$$\dot{a}_0 = -\Phi_0^a - \frac{2}{\sqrt{3}}\kappa_a\dot{\epsilon}a_1 \qquad (3.1)$$

$$\dot{a}_1 = -\Phi_1^a + \sqrt{3}\lambda_K\dot{\epsilon} - \frac{2}{\sqrt{3}}\kappa_a\dot{\epsilon}a_0, \quad \dot{a}_2 = -\Phi_2^a \qquad (3.2)$$

$$\dot{a}_3 = -\Phi_3^a + \kappa_a\dot{\epsilon}a_3, \quad \dot{a}_4 = -\Phi_4^a - \kappa_a\dot{\epsilon}a_4. \qquad (3.3)$$

Here the dimensionless extensional rate $\dot{\epsilon} = \tau_{\text{ref}}\dot{\epsilon}$ is used.

In order to compare the results with [134] the curve is scaled as follows. For the amended potential the maximum of the order parameter is $a_{\max} = 2.5$ and hence the order parameter a is scaled by a factor 2.5. On the other hand, in [134] the extensional rate is scaled by the diffusion constant D_R, while in equations (3.1-3.3) $\dot{\epsilon}$ is scaled by τ_{ref}. From the Fokker-Planck equation it is possible to derive a relation between the diffusion constant and phenomenological coefficients: $6D_R = A\tau_a^{-1}$ [44]. Finally, the temperature $\vartheta = -3.67$ was used such that the equilibrium value of the order parameter coincides with the equilibrium value of Maffettone et. al. [134].

Figure 3.1 shows the orientational order parameter as a function of the dimensionless extension rate $\dot{\epsilon}$. Very good agreement is found between the numerical solution of the Fokker-Planck equation and the alignment tensor theory with the new, amended potential for all values of the extension rate. The Landau-de Gennes potential, however, cannot be used for extension rates $\dot{\epsilon} \gtrsim 1$. In [134] the numerical values of the Fokker-Planck equations fits very well to experimental values of a four-rolls-mill flow. Hence, the amended potential is appropriate to simulate the orientational behavior subjected to extensional flows. Furthermore, the alignment tensor equation corresponds to the Fokker-Planck equation via closure approximations and therefore a reasonable closure approximation is given by the amended potential function rather than the Landau-de Gennes potential.

Similar results are be obtained for different values of the parameter κ_a as shown in Fig. 3.1. For intermediate extensional rates $\dot{\epsilon} \approx 1$, the values of the order parameter differ by roughly 10% for different choices of κ_a.

The amended potential function, for the uniaxial case ($a_{\max} = 2.5$), is compared with the Landau-de Gennes potential in Fig. (1.5, see section (1.6)). The difference of the amended potential to the Landau-de Gennes potential near the pseudo-critical temperature $\vartheta = 0$ is very small for $a \sim 1$, so that the same dynamical solutions at nearly the same model parameter Wi, λ_K for $\vartheta = 0$ can be found with a slightly shifted value of the tumbling parameter value λ_K as expressed in Fig. 3.2. The small shift can be explained by the different value of the equilibrium value of the order parameter a_{eq}. The equilibrium value is given by the condition $\Phi'(a_{\text{eq}}) = 0$ and therefore depends on the value of $a_{\max}$. For $a_{\max} = 2.5$, $a_{\text{eq}} = 0.9667$ whereas $a_{\text{eq}} = 1.0$ for the Landau-de Gennes potential.

CHAPTER 3. ORIENTATIONAL BULK DYNAMICS OF NON-POLAR HARD-ROD FLUIDS

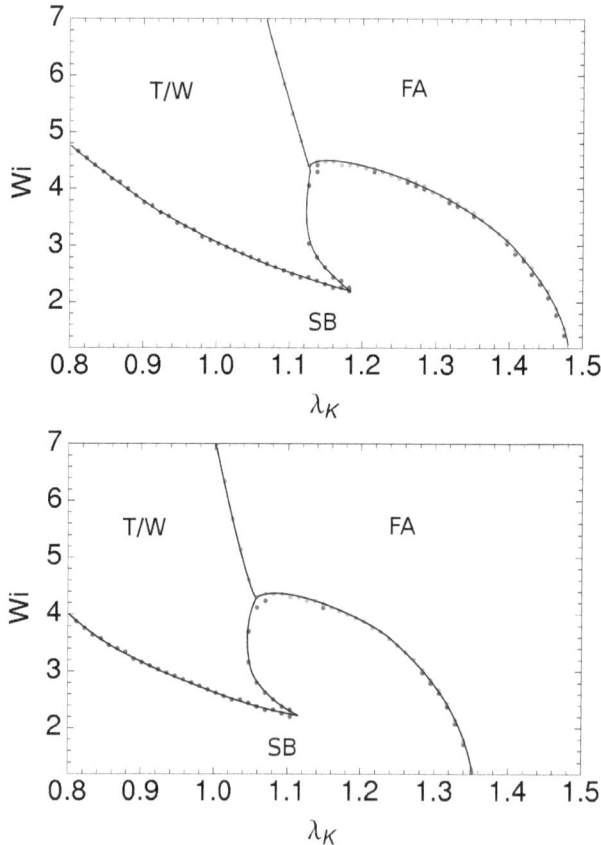

Figure 3.2: The rheological diagram is shown for flow alignment (FA), in-plane (T/W) and out-plane regimes (symmetry breaking states, SB). In the lower panel, the amended potential and in the upper panel the Landau-de Gennes potential was used. In both diagrams the reduced temperature is $\vartheta = 0$ and the parameter $\kappa_a = 0$.

3.1 Review of the Characteristic Solutions for the Orientational Dynamics

In the recent years the computer power increased and it become possible to solve the system of partial differential equations (1.84) numerically. Depending on the relevant model parameters Wi, λ_K, κ_a, ϑ, the solutions of the dynamic equations (1.84) for the spatially homogeneous alignment tensors either approach a steady state or are time-dependent when a stationary shear rate is imposed. Furthermore, solutions which, for long times, maintain the symmetry of the plane Couette type velocity gradient and where the tensor components a_3 and a_4 vanish have to be distinguished from symmetry breaking solutions where these components are non-zero. The latter ones are also referred to as 'out-of-plane' solutions, in contradistinction to the 'in-plane' states where the 'main' director, i.e., the axis associated with the largest eigenvalue of the alignment tensor is in the flow plane determined by the direction of the flow and its gradient. The following types of orbits have been found [43, 78, 79]:

(i) **Symmetry adapted solutions** with $a_3 = a_4 = 0$ which comprise
Flow aligning state (FA): This stationary in-plane state flow aligns with $a_0 < 0$ and is characteristic for high shear rates or high aspect ratios, i.e λ_K.
Tumbling solutions (T): In the in-plane tumbling state the principal director, i.e. the eigenvector corresponding to the largest eigenvalue, rotates about the vorticity axis.
Wagging solutions (W): In this state the principal director is in a in-plane wagging or librational motion about the flow direction.
Log-rolling solutions: The alignment is stationary with $a_1 = a_2 = 0$ and $a_0 > 0$. This out-of-plane solution is instable, in most cases.

The flow aligning, tumbling and wagging solutions also found in different theoretical models [21, 135–137] (for a bifurcation analysis see [42, 138–142]) and Brownian dynamics simulations [37–39]. This orientational behavior was found [32, 35, 143] in experiments of liquid crystal polymers and nano rod colloids.

(ii) **Symmetry breaking solutions** with $a_3 \neq 0, a_4 \neq 0$ (for a discussion of the symmetry see [144]), more specifically:
Stationary symmetry breaking solutions which occur in pairs of a_3, a_4 and $-a_3, -a_4$,
Kayaking-tumbling solutions (KT): In that case the projection of the principal director onto the flow plane describes a tumbling motion (see, Fig. 3.3),
Kayaking-wagging solutions (KW): The periodic orbit of the principal director projected onto the flow plane describes a wagging motion (see, Fig. 3.3).
Complex solutions (C): This state is related to complicated motion of the alignment tensor and includes periodic orbits composed of sequences of KT and KW motion with multiple periodicity as well as aperiodic, erratic orbits. The largest Lyapunov exponent for the latter orbits is positive, i.e., these orbits are *chaotic*.

A detailed overview of the solutions depending on the tumbling parameter λ_K and the Weissenberg number Wi for $\vartheta = 0$ is given in [43, 44], whereas $\dot{\gamma}$ instead of Wi and the Landau-de Gennes potential is used.

The rheological (phase) diagram is sensitive against changes of the reduced temperature (concentration) ϑ. Deep in the nematic phase ($\vartheta = -0.8$) for small and moderate aspect ratios

CHAPTER 3. ORIENTATIONAL BULK DYNAMICS OF NON-POLAR HARD-ROD FLUIDS

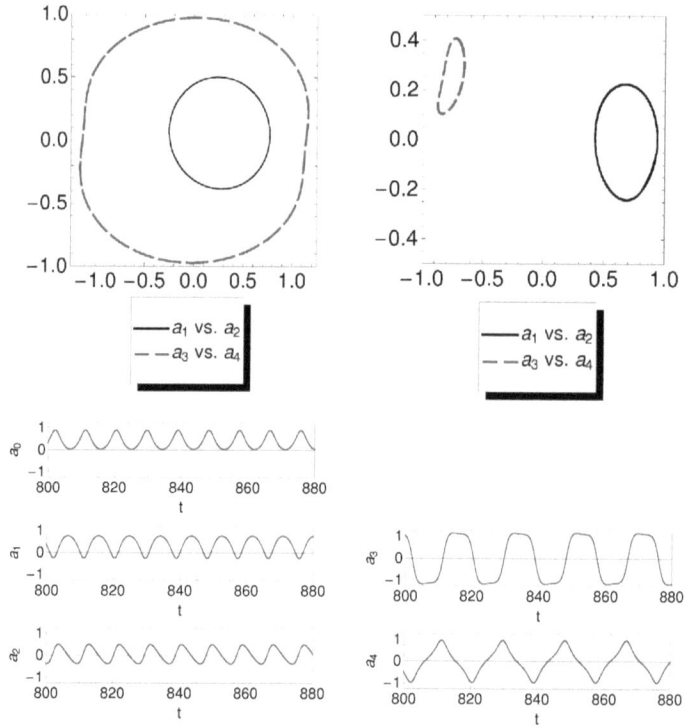

Figure 3.3: The upper panel shows the kayaking tumbling (left) and kayaking wagging (right) orbits, respectively. The time evolution for the tensor components $a_1...a_4$ for kayaking tumbling is displayed in the lower panel. The model parameters are $\lambda_K = 1$, $\kappa_a = 0$, $\vartheta = 0$ and $Wi = 1$ ($Wi = 5$ for kayaking wagging).

($\propto \lambda_{\mathrm{K}}$) of the molecules and weak shear symmetry breaking states are pronounced. High shear rates force the attractor into the shear plane and the principal director shows tumbling and wagging motion. Very long molecules (high aspect ratio) aligns in the flow. For higher model temperature ϑ the flow alignment solution is preferred. In Fig. 3.4 the rheological solution diagram for different temperatures is illustrated. Close to the isotropic-nematic phase coexistence ($\vartheta = 0.8$) symmetry breaking states occur only in very small parameter regions. For a given choice of parameters, in general, only a subset of these solutions are found by increasing the Weissenberg number Wi. The type of orientational behavior strongly affects the rheological behavior of the fluid.

In the following section, results for the orientational behavior are presented as functions of the Weissenberg number and of the strength of its fluctuating part for a few selected values of λ_{K}, Wi and the reduced temperature $\vartheta = 0$.

3.2 Robustness of Periodic and Chaotic Solutions

3.2.1 Modeling of Shear Rate Perturbations

In experiments the shear rate cannot be kept absolutely constant. It can be perturbed by the roughness of the shear plates or by the motion of the plates itself. If such fluctuations are very fast compared with the time scale of the orientational dynamics they can be neglected. Otherwise shear rate fluctuations should be taken into account in order to make the theoretical analysis more realistic.

In order to study the influence of fluctuations, time-dependent Weissenberg number of the form

$$Wi(t) = Wi_0(1 + \xi f(t)), \tag{3.4}$$

is considered. The function $f(t)$ modeling fluctuations can be an analytical function or given by numerical random numbers ($f(t_i)$). In every case $f(t)$ models Gaussian distributed random numbers with zero mean and variance $\sigma = 0.25$. Here Wi_0 is the mean value of $Wi(t)$ and ξ measures the strength or amplitude of noise. The analytical function $f(t)$ is given by

$$f(t) = \frac{1}{5}\left(g(t) + g(\frac{t}{10}) + g(\frac{t}{100}) + g(\frac{t}{\sqrt{10}}) + g(\frac{10t}{\sqrt{10}})\right) \tag{3.5}$$

as a linear combination of the function

$$g(t) = \cos(\pi \sin(\pi t))\sin(\pi(t + \sin(e\pi t))). \tag{3.6}$$

In the case where Gaussian random numbers calculated numerically the function reads

$$f(t) = \sum_{i=0}^{n} \eta_i \chi_{\Delta t}(t - i\Delta t), \tag{3.7}$$

where η_i denotes Gaussian distributed random numbers and

$$\chi_A(x) = \begin{cases} 1 & x \in A \\ 0 & \text{otherwise.} \end{cases} \tag{3.8}$$

CHAPTER 3. ORIENTATIONAL BULK DYNAMICS OF NON-POLAR HARD-ROD FLUIDS

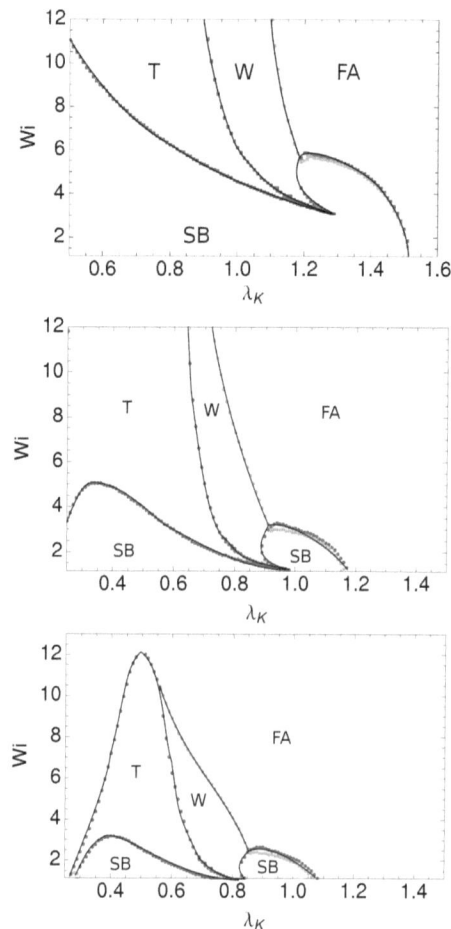

Figure 3.4: The dependence of the rheological solution diagram on the change of the reduced temperature ϑ for $\kappa_a = 0$ is given. From top to bottom: $\vartheta = -0.8$, $\vartheta = 0.6$, $\vartheta = 0.8$.

3.2. ROBUSTNESS OF PERIODIC AND CHAOTIC SOLUTIONS

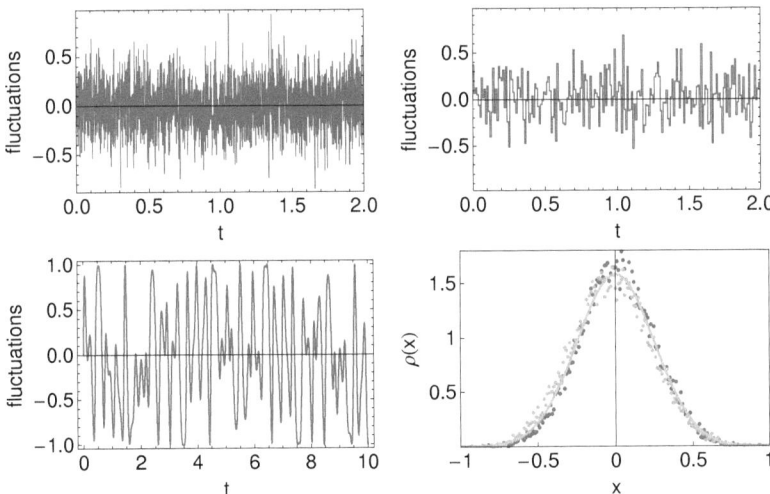

Figure 3.5: The upper panel show random fluctuations calculated numerically at different time scales. On the left the function is constant for $\triangle t = 0.001$ and on the right for $\triangle t = 0.01$. Pseudo-random fluctuation given by $f(t)$ as a nonlinear function of the time t, Eq. (3.5), in the lower left panel. The distribution function $p(x)$ of dimensionless pseudo-random and random (blue points) fluctuations x is displayed on the right hand side.

CHAPTER 3. ORIENTATIONAL BULK DYNAMICS OF NON-POLAR HARD-ROD FLUIDS

In Fig. 3.5 the model function $f(t)$ is plotted against the time t. The extremal values for f are ± 1. In the lower right panel of the figure the distribution are compared with the Gauss-function $\left[\sqrt{2\pi}\sigma\right]^{-1} \exp\left[-\frac{x^2}{2\sigma^2}\right]$, where $\sigma = 0.25$ is the square root of the second cumulant.

3.2.2 Isotropic Phase, Flow Alignment and Periodic Solutions

In the isotropic regime, every aligned initial configuration decays to a state with random orientation such that the components of the alignment tensor tend to zero. In order to investigate the effect of shear rate fluctuations on this behavior, the shear rate is set to $Wi(t) = \xi f(t)$. No significant qualitative effects of fluctuating shear rates on the dynamical solutions are found. In Fig. 3.6, the time evolution of the scalar order parameter $a = \sqrt{\mathbf{a}:\mathbf{a}}$ is shown both, in the absence of noise and in the presence of small and strong shear rate fluctuations. In the nematic phase the orientational flow leads to enhanced alignment for high values of Wi and λ_K. The molecules show mostly in one preferred direction. Therefore, after a transient period, the components of the alignment tensor reach stationary values.

In principle the solutions with noisy shear rates are not different from the usual solutions. Only some fluctuations on the stationary values of alignment tensor components can be observed. This means that the molecules fluctuate around the preferred direction. In Fig. 3.6, for example, the time evolution of the alignment tensor component a_1 is shown for $\xi = 0$ and $\xi = 5.0$.

Periodic solutions occur in a manifold way depending on the model parameter λ_K and Wi, as was discussed in Sec.3.1. For example, for the parameters $Wi_0 = 2.0$ and $\lambda_K = 0.9$, show kayaking tumbling (KT) solution whereas for $\dot{\gamma}_0 = 5.0$ and $\lambda_K = 0.9$ tumbling (T) solution is observed [43, 78]. One could suspect that strong fluctuations of the shear rate Wi lead to a transition of the KT to the T solution. But even for the strongest fluctuations investigated the KT solutions for $Wi_0 = 2$ persists, despite the fact that the largest values of the shear rates fluctuating around $Wi_0 = 2$ are in a range where a tumbling solution occurs for stationary shear. In Fig. 3.7, orbits of non-fluctuating shear rates are compared with orbits obtained from strong fluctuations ($\xi = 50, \xi = 170$). Again, the trajectory is perturbed by fluctuations but the character of the solution is conserved. Only at very strong, physically unrealistic, fluctuations the KT solution changes to a noisy wagging like solution.

The robustness of solutions against fluctuations is also observed for other periodic solutions. In Fig. 3.8, the phase plots of wagging solutions for several values of ξ is shown. It can be seen that for increasing ξ the phase cycle is smeared out because the rapid distorsion seen in the solution overlaps the original curves. In principle all orbits show the same response to small and intermediate fluctuations. Therefore, it can be concluded that fluctuations perturb the trajectory but the system tries to follow the original dynamic without distorsions Only at very high values of the parameter ξ, the dynamical behavior is described by the character of the fluctuations and differences between numerical and analytical calculated noise becomes significant.

3.2. ROBUSTNESS OF PERIODIC AND CHAOTIC SOLUTIONS

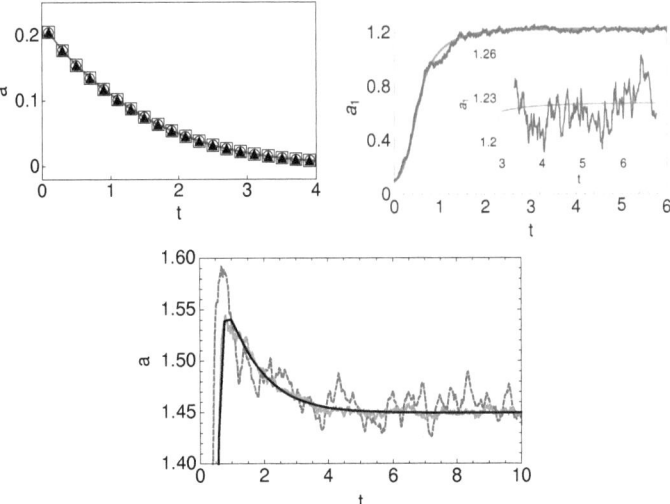

Figure 3.6: Upper left Fig.: The scalar order parameter a in the isotropic regime as a function of the time is given for several values of the parameter ξ (cubic for $\xi = 0$, triangle for $\xi = 5$ and circle for $\xi = 10$). The reduced temperature is $\vartheta = 1$; the model parameters are $\lambda_K = 1.0$, Wi_0 and $\kappa_a = 0$. The time evolution of the tensor component a_1 in the flow alignment regime for strong fluctuations ($\xi = 5.0$) of the shear rate around $Wi_0 = 2.0$ and in the absence of fluctuations (gray line) is compared (right figure). The reduced temperature is $\vartheta = 0$; the model parameters are $\lambda_K = 1.5$, $\kappa_a = 0$ and $\triangle t = 0.001$ (left figure). Lower Fig.: The time evolution of the scalar order parameter a in the flow alignment regime with fluctuations of the shear rate ($\xi = 2.5$) around $Wi_0 = 2.0$ calculated numerically ($\triangle t = 0.001$) and analytically (dashed line) compared with the original solution (gray line, $\xi = 0$).

CHAPTER 3. ORIENTATIONAL BULK DYNAMICS OF NON-POLAR HARD-ROD FLUIDS

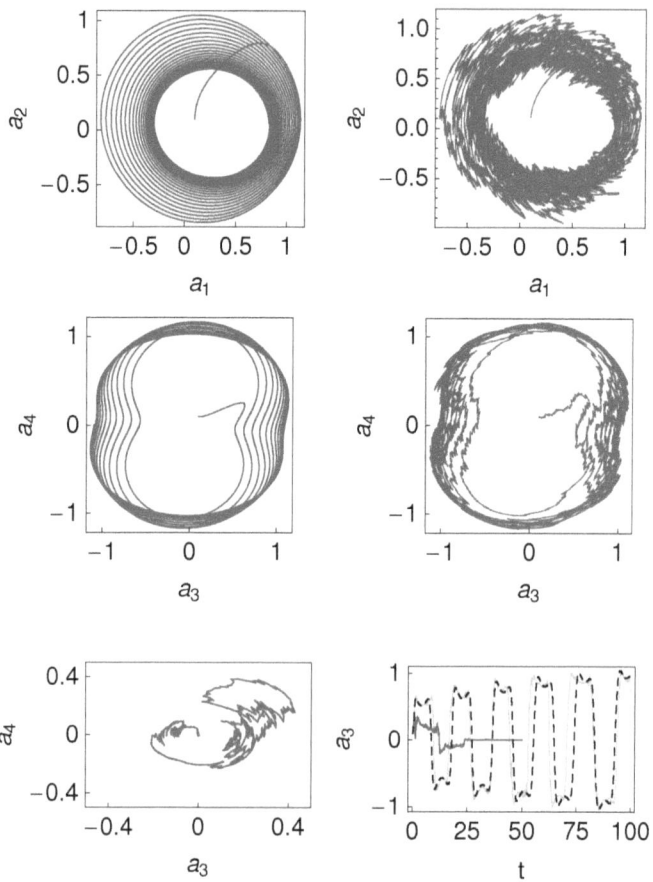

Figure 3.7: The orbits a_4 versus a_3, a_2 versus a_1 and the time evolution of a_3 is plotted for several coupling constants: $\xi = 0$ (upper left), $\xi = 50$ (upper right) and $\xi = 170$ (lower left). In the lower right the graphs for $\xi = 0$ (dashed line), $\xi = 50$ and $\xi = 170$ (gray line) are displayed. The temperature is $\vartheta = 0$; the model parameters are $\lambda_K = 1.0$, $Wi_0 = 1.0$, $\kappa_a = 0$ and $\triangle t = 0.001$.

3.2. ROBUSTNESS OF PERIODIC AND CHAOTIC SOLUTIONS

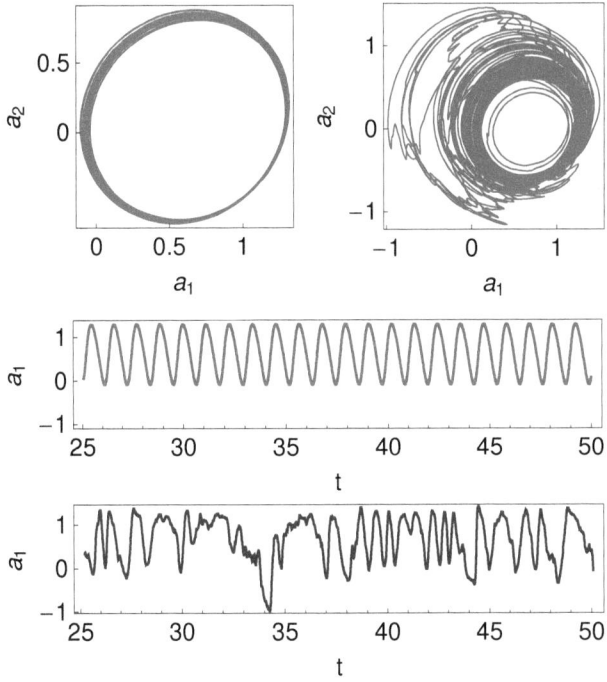

Figure 3.8: The orbits a_1 versus a_2 in the tumbling regime for $\xi = 1$ upper left and $\xi = 5$ upper right. In the middle the time evolution of a_1 for $\xi = 0$ and $\xi = 5$ (lower) is given. The reduced temperature is $\vartheta = 0$; the model parameters are $\lambda_K = 0.8$, $Wi = 6$, $\kappa_a = 0$ and $\triangle t = 0.001$.

CHAPTER 3. ORIENTATIONAL BULK DYNAMICS OF NON-POLAR HARD-ROD FLUIDS

Figure 3.9: The Figure sketch the calculation of the largest Lyapunov exponent. The solution trajectory is perturbed by the vector d_0. The difference of the perturbed trajectory and the original once after some time units is given by the vector d_1. The vector d_1 is scaled to the length of d_0 and the procedure starts again. The vectors d_k are used to determine the largest Lyapunov exponent, i.e. $\Lambda = \frac{1}{N}\sum_{k=1}^{N} \ln \frac{d_k}{d_0}$.

3.2.3 Chaotic Solutions

It seems to be obvious that fluctuations effect chaotic solutions because of its sensitivity on changing the model parameter Wi or λ_K. It is astonishing that chaotic solutions are very robust against uncorrelated noise. In order to investigate the effect of shear rate fluctuations on the chaotic solutions the largest Lyapunov exponent is calculated. For that purpose the trajectories of the alignment tensor components are perturbed by a small vector d_0. After some time-steps the distance between the perturbed and unperturbed trajectory is measured by d_1 and the perturbed trajectory is rescaled such that the new distance is d_0. The logarithm of the ratio d_1/d_0 is calculated. This procedure is repeated for many steps and the average of the logarithms leads to the largest Lyapunov exponent (see, Fig. 3.9). The numerical implementation of the algorithm is described in [145].

In Fig. 3.10 the largest Lyapunov exponent against the fluctuation amplitude is displayed.

The Lyapunov exponent is positive for a wide range of the parameter ξ. That means for fluctuations up to thirteen times higher than the averaged Weissenberg number (shear rate) the solution is still chaotic. Such high fluctuations are not realistic for experiments as long as the flow is laminar. Therefore chaotic behavior is not sensitive against uncorrelated fluctuations of the shear rate in experiments.

The same could be observed for fluctuations that where generated by the analytical function $f(t)$ as shown in Fig. 3.10. Again fluctuations up to fourteen times higher than Wi_0 not perturb the chaotic states. Despite the similarities there is a significant difference between the two plots. In the range between $\xi = 0$ and $\xi = 6$ the Lyapunov exponent increases in the case of numerical calculated fluctuations whereas if the fluctuations modeled by the analytic function $f(t)$ it decreases. This can be explained by the fact that the Lyapunov exponent measure the noise if it is chaotic like the fluctuations calculated numerically.

Also the trajectories of the alignment tensor components are not changed. Fig. 3.11 shows that the trajectory in the phase space is practically not affected even by rather strong fluctuations. The trajectory is only modulated by some noisy distorsions as already noticed for periodic solutions.

3.2. ROBUSTNESS OF PERIODIC AND CHAOTIC SOLUTIONS

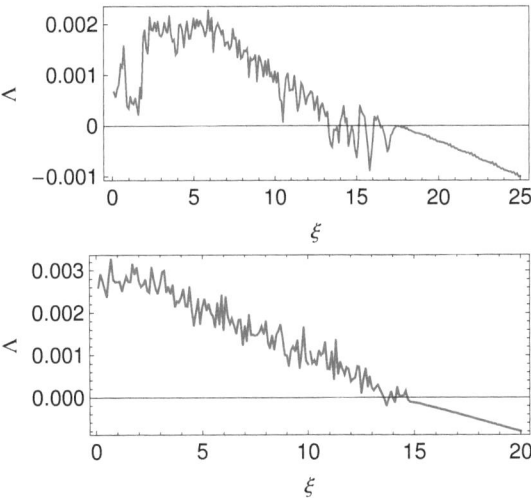

Figure 3.10: The largest Lyapunov exponent is plotted against the amplitude of fluctuations ξ. Fluctuations are modeled by Gaussian random numbers. The reduced temperature is $\vartheta = 0$; the model parameters are $\lambda_{\text{K}} = 1.17$, $Wi_0 = 3.72$ and $\kappa_{\text{a}} = 0$. Lower: The largest Lyapunov exponent is plotted against the amplitude of fluctuations ξ, where fluctuations are modeled by Eq. (3.5).

CHAPTER 3. ORIENTATIONAL BULK DYNAMICS OF NON-POLAR HARD-ROD FLUIDS

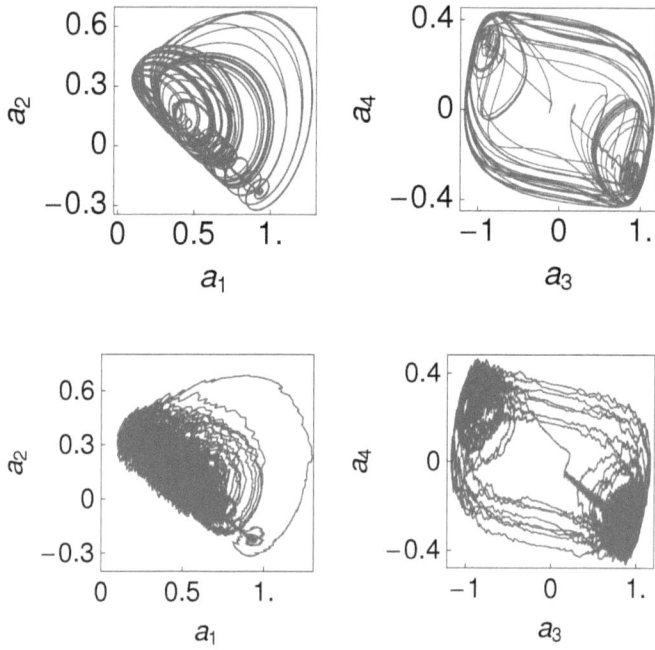

Figure 3.11: The orbit a_1 versus a_2 (left) and a_3 versus a_4 (right) in the chaotic regime at $Wi_0 = 3.7$ for several coupling constants ξ (upper Fig. $\xi = 0.05$ and lower Fig. $\xi = 8$). The reduced temperature is $\vartheta = 0$; the model parameters are $\lambda_K = 1.15$, $\kappa_a = 0$ and $\triangle t = 0.001$

3.2. ROBUSTNESS OF PERIODIC AND CHAOTIC SOLUTIONS

It is surprising that chaotic solutions are very sensitive against the change of averaged shear rate (Weissenberg number) but not on fluctuations. The study of the largest Lyapunov exponent shows that no small parameter values ξ exist such that the chaotic behavior is lost. In the case of periodic solutions it was shown that for increasing fluctuation strength the phase trajectory is smeared out more and more. The orientation of nematic follows the original phase trajectory. Small fluctuations indicate oscillations of the orientation around the original trajectory. In conclusion fluctuations of the shear rate do not affect the dynamics of hard rods in the shear flow.

This result may change when the time scale of fluctuations is in the order of orientational relaxation times, ie. higher $\triangle t$. In the Fig. 3.12 perturbed orbits for different $\triangle t$ are compared to unperturbed orbits. For $\triangle t = 0.001$ and $\triangle t = 0.01$ the trajectory is closed to the unperturbed orbit. This is changed for long time fluctuations $\triangle t = 0.1$. The orbit strongly smears out. However to indicate the difference unphysical high fluctuation strength ($\xi = 10$) are necessary.

The robustness of chaotic solutions can be extended to mixed shear and extensional flows. If the plane Couette flow is modified by a planar straining flow the chaotic behavior persists up to a threshold straining strength. Moreover the straining component can induce chaos from periodic shear response [146].

However, when the velocity gradient fluctuates itself the system is less robust. In [147] Leal *et. al.* could show that a variation of the velocity gradient and its effect on other flow components may destroy the periodic director behavior.

CHAPTER 3. ORIENTATIONAL BULK DYNAMICS OF NON-POLAR HARD-ROD FLUIDS

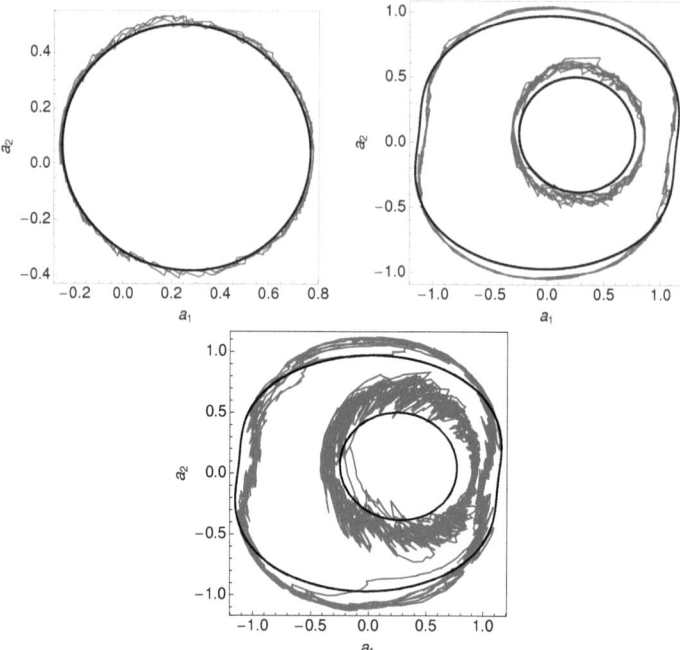

Figure 3.12: Orbits of the tensorial components for different time scales of fluctuations. From left to right $\triangle t = 0.001$, $\triangle t = 0.01$, and $\triangle t = 0.1$. The parameter values are $\xi = 10$, $Wi = 1$, $\vartheta = 0$, $\lambda_{\mathrm{K}} = 1$.

4

Spatially Inhomogeneous Dynamics of Non-Polar Hard-Rod Fluids

In the previous chapter the characteristic solutions and their robustness for homogeneous systems with a constant shear rate was discussed. In this section the orientational behavior of hard-rod fluids between two plates (one-dimensional confinement) is investigated. First, the order parameter profile in equilibrium is briefly presented. Second, boundary conditions within the framework of irreversible thermodynamics are derived under the assumption that the wall is given by a solid-fluid interface. In can be shown that this boundary conditions yield an apparent velocity slip. In section (4.3.1) the orientational behavior and rheological diagrams for the hard-rod fluid subjected to an imposed shear flow is presented. Finally, the full hydrodynamic equations of the non-polar hard-rod fluid are solved for the full alignment tensor in one spatial dimension. The non-Newtonian flow feedback reveals to a surprising local spurt of the velocity related to the orientational spatio-temporal mesostructure. The flow-orientation relation is discussed and parameter dependence are given. It turns out that the inherent mesoscopic structure is determined by the Ericksen number.

4.1 Equilibrium States

In spatially heterogeneous systems boundary conditions affect significantly the equilibrium and non-equilibrium behavior of the fluid. In equilibrium, that is zero flow gradients and stationarity, and uniaxial alignment the equations (1.84) reduce to

$$\frac{d^2 a}{dy^2} = \frac{Er}{De}\Phi^a(a,\vartheta). \tag{4.1}$$

This is a nonlinear second order ordinary differential equation. Multiplying Eq. (4.1) with $\frac{da}{dy}$ and integrating gives

$$\left(\frac{da}{dy}\right)^2 = 2\int \Phi^a(a)da + c, \tag{4.2}$$

where c is a arbitrary integration constant. For simplicity the derivative of the Landau-de Gennes potential Φ^a is used. The solution is given by

$$y(a) = \int \frac{da}{\sqrt{2\frac{Er}{De}}\sqrt{\vartheta a^2 - 2a^3 + a^4 + c}} \tag{4.3}$$

CHAPTER 4. SPATIALLY INHOMOGENEOUS DYNAMICS OF NON-POLAR HARD-ROD FLUIDS

In the nematic phase the potential Φ can be approximated by

$$\begin{aligned}\Phi_{app} &= \Phi(a_{eq}) + \Phi(a_{eq})'(a-a_{eq}) + \frac{1}{2}\Phi(a_{eq})''(a-a_{eq})^2 + O(a^3) \\ &= -\frac{27}{32} + \frac{9}{4}\left(a-\frac{3}{2}\right)^2 + O(a^3)\end{aligned} \qquad (4.4)$$

The integral (4.3) simplifies to

$$y(a) = \int \frac{da}{\sqrt{2\frac{Er}{De}}\sqrt{-\frac{27}{16} + \frac{9}{2}\left(a - \frac{3}{2}\right)^2 + c}} \qquad (4.5)$$

Integration and rearranging the equation yields

$$a(y) = \frac{1}{24}\left(e^{-\frac{3(d-y)}{\sqrt{2}\ell}} + 54e^{\frac{3(d-y)}{\sqrt{2}\ell}} - 64e^{-\frac{3(d-y)}{\sqrt{2}\ell}}c\right), \qquad (4.6)$$

where $\ell_c = \sqrt{\frac{Er}{De}}$ and (c,d) are integration constants. If the order parameter a vanishes at the boundary, i.e. $a(-1) = 0$, $a(1) = 0$ then the $a(y)$ simplifies to

$$a(y) = \frac{3}{2}\left(1 - \frac{\cosh\left(\frac{3y}{\sqrt{2}\ell}\right)}{\cosh\left(\frac{3}{\sqrt{2}\ell}\right)}\right). \qquad (4.7)$$

In Fig. 4.1 the order parameter profile between two plates is shown. For wider plate separations (full lines) the order parameter increases rapidly to the equilibrium value of 3/2. For narrow separation (dashed lines) the influence of the wall strong, such that the equilibrium value is not reached. The value of the order parameter in the middle of the gap depends on the value of ℓ_c. In Fig 4.1 the order parameter at $y = 0$ versus the ℓ_c is displayed. Later it will be shown that the parameter ℓ_c determine the formation of structure of sheared rod.

In the next section appropriate boundary conditions for the alignment flux are derived in analogy to the thermo-hydrodynamics and the theory of rarefied gases and consequences are discussed.

4.2 Apparent Slip of the Isotropic State Subjected to a Flow

The equations of thermo-hydrodynamics, based on the local conservation laws and on simple constitutive relations for the heat flux and the viscous pressure tensor, have to be supplemented by boundary conditions. Temperature jump and velocity slip boundary conditions have been proposed over a century ago in order to describe boundary and surface effects in *rarefied gases* [91] where the mean free path ℓ of the molecules can become comparable with the relevant macroscopic lengths. For dense fluids, similar effects have to be taken into consideration in *microfluidics* and *nano-rheology* [148–152] where the size of the molecules is no longer extremely small compared with macroscopic length scales. Especially, the apparent

4.2. APPARENT SLIP OF THE ISOTROPIC STATE SUBJECTED TO A FLOW

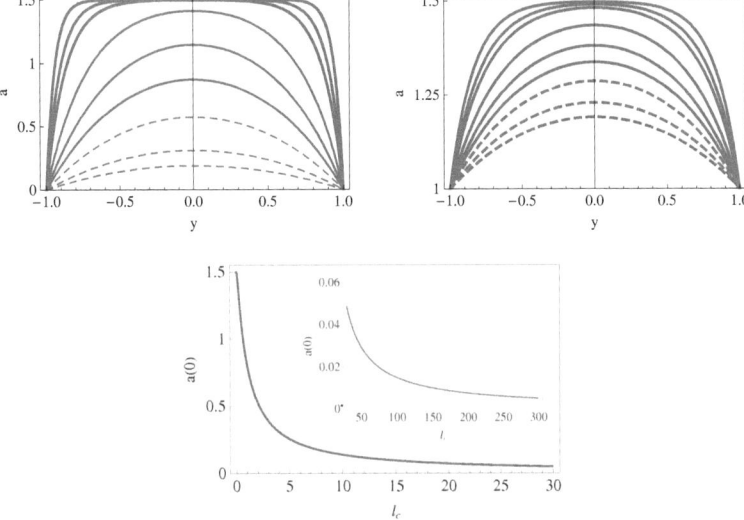

Figure 4.1: The order parameter profile in the gap is shown for different parameter values of ℓ_c (from the dashed lines to the thick lines, 0.12, 0.2, 0.28, 0.6, 1.0, 1.4, 2, 3, 4). In the left figure $a(-1) = a(1) = 0$ and in the right figure $a(-1) = a(1) = 1$, respectively. The dependence of the order parameter at $y = 0$ vs. the parameter ℓ_c. is displayed (lower panel).

CHAPTER 4. SPATIALLY INHOMOGENEOUS DYNAMICS OF NON-POLAR HARD-ROD FLUIDS

slip caused by the molecular interaction with the solid surface was the reason for many theoretical and experimental studies, see for eg. [71–76]. For fluids composed of particles with orientational degrees of freedom, additional constitutive equations govern the dynamics of the molecular alignment and again, boundary conditions are needed for spatially inhomogeneous situations.

Here, boundary conditions are formulated for the second rank alignment tensor describing the orientation of hard rods and for the velocity slip. The guiding principle, in the spirit of irreversible thermodynamics, is the same as that originally suggested for gases [80] viz.: i) the entropy production at an interface is inferred from the entropy flux in the bulk fluid, ii) the boundary conditions are set up such that the interfacial entropy production is positive definite. The extension to molecular gas and to molecular liquids was presented in [81, 82]. For a special case meant for isothermal flow of molecular liquids, polymeric melts and nematic liquid crystals in the isotropic phase, it is demonstrated that the coupling between the alignment tensor and the friction pressure tensor which underlies the flow birefringence and shear-thinning leads to an apparent velocity slip even when the velocity obeys a stick boundary condition. The velocity and alignment profiles, as well as the effective viscosities are calculated for plane and cylindrical Couette and plane Poiseuille flow, as well as the flow down an inclined plane. The dependence of these quantities and of the apparent slip velocity on a microscopic length parameter and on the ratio between the first and second Newtonian viscosities are discussed. In experiments slip lengths and the viscosities of thin films of Newtonian liquids were measured and studied by Jacobs *et. al.* [83]. Furthermore, a recent thermodynamic formulation of boundary conditions building upon the pioneering work of Waldmann [80] was derived in [84–86].

4.2.1 Boundary Conditions

In the spirit of irreversible thermodynamics, L. Waldmann [80] suggested to set up boundary conditions such that the interfacial entropy production is positive definite. For a fluid in contact with a solid wall moving with the velocity $\mathbf{v}^w$ one has [82, 153]

$$\theta^w = \int df \, T^{-1}[k_\mu^{\tan}(v_\mu - v_\mu^w)^{\tan} + (\rho/m)k_B n_\lambda b_{\lambda\mu\nu}\Phi_{\mu\nu}]. \tag{4.8}$$

Here df is the surface element, $\mathbf{n}$ is the outer normal of the fluid and the superscript $...^{\tan}$ indicates a tangential component which is parallel to the surface, e.g. $v_\mu^{\tan} = v_\mu - n_\mu n_\nu v_\nu$. Furthermore,

$$k_\mu = n_\nu(P_{\nu\mu} - P\delta_{\nu\mu}) \tag{4.9}$$

is the friction force density. It is tacitly assumed that P is the hydrostatic equilibrium pressure, which means that a scalar pressure associated with the bulk viscosity is disregarded or that the fluid is practically incompressible, viz.: $\nabla \cdot \mathbf{v} = 0$.

Just as in [82], the notation is followed which H. Vestner [81] used for molecular gases where the equations are linear in the alignment tensor which, in addition, has a different

4.2. APPARENT SLIP OF THE ISOTROPIC STATE SUBJECTED TO A FLOW

microscopic meaning. The boundary conditions for the velocity and the alignment tensor are:

$$(v_\mu - v_\mu^{\text{w}})^{\text{tan}} = C_{\text{m}}\, v_{\text{th}} p_{\text{kin}}^{-1} k_\mu^{\text{tan}} + C_{\text{ma}}\, n_\lambda b_{\lambda\mu\nu} n_\nu, \tag{4.10}$$

$$\Phi_{\mu\nu}^{a} = C_{\text{am}}\, p_{\text{kin}}^{-1} \overline{k_\mu^{\text{tan}} n_\nu} + C_{\text{a}}\, v_{\text{th}}^{-1} n_\lambda b_{\lambda\mu\nu}. \tag{4.11}$$

Here the kinetic pressure $p_{\text{kin}} = (\rho/m)k_B T$ and the thermal velocity $v_{\text{th}} = \sqrt{k_B T/m}$ are used as reference values for the pressure and the velocity. The dimensionless coefficients $C_{..}$ specify the boundary behavior, the subscripts m and a refer to *mechanical* and *alignment*, respectively. The diagonal coefficients are non-negative, viz,: $C_{\text{m}} \geq 0$, $C_{\text{a}} \geq 0$. The off-diagonal ones obey the Onsager-Casimir relation $C_{\text{am}} = -C_{\text{ma}}$. The slip velocity is $\delta v_\mu = -(v_\mu - v_\mu^{\text{w}})^{\text{tan}}$. For $C_{\text{ma}} = 0$ and with $k_\mu = -\eta\, n_\nu \nabla_\nu v_\mu$, where η is the shear viscosity, (4.10) is equivalent to

$$\delta v_\mu = \ell_v\, n_\nu \nabla_\nu v_\mu^{\text{tan}}, \tag{4.12}$$

with the slip length

$$\ell_v = C_{\text{m}}\, \eta\, v_{\text{th}} p_{\text{kin}}^{-1} \geq 0, \tag{4.13}$$

In the following, it will be demonstrated that the coupling between the alignment and the flow field leads to an apparent velocity slip even when $C_{\text{m}} = 0$ and consequently $\ell_v = 0$.

4.2.2 Isotropic Phase and Small Shear Rates

Next the equations for the bulk fluid and the boundary conditions are applied to special geometries where the spatial dependence is essentially one-dimensional, viz.: to a plane Couette and a plane Poiseuille flow. The attention is focused on the isotropic phase where terms non-linear in the alignment tensor can be disregarded and on the Newtonian flow regime where only terms linear in the velocity gradient are taken into account. The antisymmetric part of the pressure tensor vanishes in this case. Then, with (1.47), (1.48) and (1.49), the momentum balance equation (1.46) reduces to

$$\rho \frac{\partial v_\mu}{\partial t} + \nabla_\lambda P_{\lambda\mu} = \eta_{\text{iso}} \Delta v_\mu - p_{\text{kin}} \sqrt{2}\, \frac{\tau_{\text{ap}}}{\tau_{\text{a}}}\, \varphi_\mu, \tag{4.14}$$

with

$$\varphi_\mu = \nabla_\nu \Phi_{\nu\mu}^a. \tag{4.15}$$

Similarly, the relaxation equation (1.43) for the alignment tensor is approximated by

$$\frac{\partial a_{\mu\nu}}{\partial t} + \tau_{\text{a}}^{-1} \left(\Phi_{\mu\nu}^a - \ell_{\text{a}}^2 \Delta \Phi_{\mu\nu}^a \right) = -\sqrt{2}\, \frac{\tau_{\text{ap}}}{\tau_{\text{a}}}\, \overline{\nabla_\nu v_\mu}, \tag{4.16}$$

and it is understood that only terms linear in the alignment tensor are considered in $\Phi_{\mu\nu}^a$, i.e. $\Phi_{\mu\nu}^a$. The characteristic length ℓ_a associated with the alignment diffusion is defined by

$$\ell_{\text{a}}^2 = D_{\text{a}} \tau_{\text{a}}. \tag{4.17}$$

For stationary situation, application of ∇_ν on (4.16) leads to

$$\left(\varphi_\mu - \ell_{\text{a}}^2 \Delta \varphi_\mu \right) = -(1/2)\sqrt{2}\, \tau_{\text{ap}}\, \Delta v_\mu. \tag{4.18}$$

CHAPTER 4. SPATIALLY INHOMOGENEOUS DYNAMICS OF NON-POLAR HARD-ROD FLUIDS

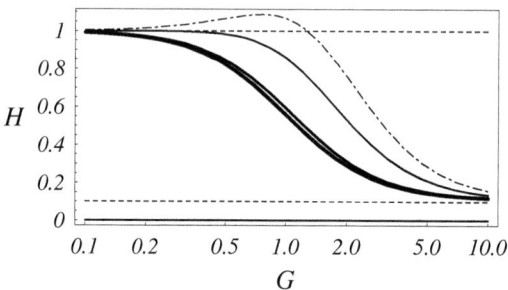

Figure 4.2: Non-Newtonian viscosity for the plane Couette flow in the isotropic phase for $Q = 8$. Here $G = \gamma A^{-1}\tau_a$ is the shear rate and H is the viscosity in units of the first Newtonian viscosity η. The values for κ are, from left to right, 0, 0.4, 1.0 and 1.2 (dashed curve).

Elimination of $\Delta \mathbf{v}$ from (4.18) with the help of (4.14), again for a steady state where the time derivatives vanish, yields

$$\left(1 + \tau_{\mathrm{ap}}^2/(\tau_0 \tau_\mathrm{a})\right) \varphi_\mu - \ell_\mathrm{a}^2 \Delta \varphi_\mu = -\eta_{\mathrm{iso}}^{-1} \tau_{\mathrm{ap}} \nabla_\lambda P_{\lambda\mu} \,. \qquad (4.19)$$

The relaxation time τ_0 is related to the second Newtonian viscosity η_{iso} by $\eta_{\mathrm{iso}} = p_{\mathrm{kin}} \tau_0$. Similarly, the first Newtonian viscosity η is linked with the relaxation time τ_p by $\eta = p_{\mathrm{kin}} \tau_\mathrm{p}$, and one has [91] $\tau_0/\tau_\mathrm{p} = 1 - \tau_{\mathrm{ap}}^2/(\tau_\mathrm{a}\tau_\mathrm{p})$. The abbreviation

$$Q = \frac{\tau_{\mathrm{ap}}^2}{\tau_0 \tau_\mathrm{a}} = \frac{\tau_{\mathrm{ap}}^2}{\tau_\mathrm{a}\tau_\mathrm{p}} \left(1 - \frac{\tau_{\mathrm{ap}}^2}{\tau_\mathrm{a}\tau_\mathrm{p}}\right)^{-1} \geq 0\,, \qquad (4.20)$$

is introduced. Notice that $\tau_{\mathrm{ap}}^2/(\tau_\mathrm{a}\tau_\mathrm{p}) < 1$. The quantity Q which is a measure for the strength of the coupling between the pressure tensor and the alignment is linked the ratio η/η_{iso} between the first and second Newtonian viscosities by

$$Q = \frac{\eta}{\eta_{\mathrm{iso}}} - 1\,. \qquad (4.21)$$

Now (4.19) is rewritten as

$$\varphi_\mu - \ell^2 \Delta \varphi_\mu = -\eta^{-1} \tau_{\mathrm{ap}} \nabla_\lambda P_{\lambda\mu} \qquad (4.22)$$

with the characteristic length ℓ related to ℓ_a by

$$\ell^2 = \ell_\mathrm{a}^2 \left(1 + Q\right)^{-1} = \ell_\mathrm{a}^2 \left(1 - \frac{\tau_{\mathrm{ap}}^2}{\tau_\mathrm{a}\tau_\mathrm{p}}\right)\,. \qquad (4.23)$$

The solutions of the homogeneous part of (4.22) couple with the velocity field due to (4.18) and via the boundary condition. Next, this point is discussed for a simple one-dimensional

4.2. APPARENT SLIP OF THE ISOTROPIC STATE SUBJECTED TO A FLOW

spatial dependence as encountered in plane Couette and in a plane Poiseuille flow. The parameter Q can be inferred from the nonlinear flow behavior, in particular from the non-Newtonian viscosity in the limits of small and large shear rates. The stationary solution of (1.43) in the isotropic phase and for a plane Couette flow where the boundary effects to be discussed below are ignored, were presented in [47]. A representative example is shown in Fig. 4.2. The transition from the first to the second Newtonian viscosity depends on the parameter κ, the limiting values determining Q are not affected by κ. The following calculations focus on the effect of the boundary conditions in the limiting case of small shear rates, viz. the linear flow regime.

4.2.3 One-Dimensional Spatial Dependence

For a flow in x-direction and its gradient in y-direction as it occurs between flat plates which are perpendicular to the y-direction one has

$$\Gamma_{\nu\mu} = \overline{\nabla_\nu v_\mu} = \gamma(y)\,\overline{e^x_\nu e^y_\mu}, \tag{4.24}$$

with the shear rate $\gamma = \gamma(y)$, the unit vectors parallel to the coordinate axes are denoted by $\mathbf{e}^x$, $\mathbf{e}^y$ and $\mathbf{e}^z$. Similarly, the ansatz

$$\Phi^a_{\mu\nu} = \sqrt{2}\,a(y)\,\overline{e^x_\nu e^y_\mu}, \tag{4.25}$$

is made which implies $\varphi_\mu = (1/2)\sqrt{2}\,a'(y)\,e^x_\mu$. The prime indicates the derivative with respect to y. In this special case and for stationary situation, (4.16) reduces to

$$a - \ell_a^2\,a'' = -\tau_{ap}\gamma. \tag{4.26}$$

Similarly, (4.22) where the momentum balance has been taken into account, is equivalent to

$$a' - \ell^2\,a''' = -\tau_{ap}\,\eta^{-1}\frac{\delta P}{L}. \tag{4.27}$$

For a Poiseuille flow the pressure gradient in x-direction is given by the ratio of the pressure difference δP and the length L of the flow device. In the case of a Couette flow one has $\delta P = 0$.

The boundary condition (4.11) for the alignment, at a wall with the outer normal in the positive y-direction now reduces to

$$a = -C_a\,v_{th}^{-1}D_a\,a' = -c_a\,\ell\,a'. \tag{4.28}$$

Here the constitutive Eq. (1.45) and $C_{am} = 0$ were used. The abbreviation

$$c_a = C_a\frac{D_a}{\ell\,v_{th}} = C_a\frac{\ell_a\sqrt{1+Q}}{\tau_a\,v_{th}} \tag{4.29}$$

was introduced. Next, solutions of the differential equations (4.26) and (4.27), with the appropriate boundary conditions are presented for the Couette and the Poiseuille flow geometry.

CHAPTER 4. SPATIALLY INHOMOGENEOUS DYNAMICS OF NON-POLAR HARD-ROD FLUIDS

4.2.4 Plane Couette Flow

A Couette flow between (identical) plates separated by the distance $2h$ is considered. The plates located at $y = h$ and $y = -h$ move with the velocities $\mathbf{u}^w$ and $-\mathbf{u}^w$, respectively, in x-direction. Here one has $\delta P = 0$ and the solution of (4.27), with the the symmetry of the set up taken into account, is

$$\alpha = \alpha_0 + \alpha_1 \cosh(y/\ell). \tag{4.30}$$

Likewise, for the shear rate the ansatz

$$\gamma = \gamma_0 + \gamma_1 \cosh(y/\ell) \tag{4.31}$$

is made. For the x-component of the velocity one has

$$u(y) = \gamma_0 y + \gamma_1 \ell \sinh(y/\ell). \tag{4.32}$$

The coefficients $\alpha_0, \alpha_1, \gamma_0, \gamma_1$ have to be determined with the help of (4.26) and of the boundary conditions.

Now a no-slip boundary condition corresponding to $C_m = C_{ma} = 0$ is assumed, cf. (4.10). This implies $u^w = \gamma_0 h + \gamma_1 \ell \sinh(h/\ell)$. The 'external' shear rate γ^{ext} is related to γ_0, γ_1 by

$$\gamma^{\text{ext}} = u^w/h = \gamma_0 + \gamma_1 (\ell/h) \sinh(h/\ell). \tag{4.33}$$

The boundary condition (4.28) for the alignment leads to

$$\alpha_0 = -\alpha_1 \left(\cosh(h/\ell) + c_a \sinh(h/\ell) \right). \tag{4.34}$$

Two further relations are needed for the determination of the coefficients, viz.

$$\alpha_0 = -\tau_{\text{ap}} \gamma_0, \quad Q \alpha_1 = \tau_{\text{ap}} \gamma_1 \tag{4.35}$$

follow from the differential equation (4.26). Insertion of these relations into (4.34) yields

$$Q \gamma_0 = \gamma_1 \left(\cosh(h/\ell) + c_a \sinh(h/\ell) \right). \tag{4.36}$$

From this relation and (4.33) follows

$$\gamma_0 = R \gamma^{\text{ext}}, \quad R = \left(1 + Q \frac{\ell}{h} \frac{\tanh(h/\ell)}{1 + c_a \tanh(h/\ell)} \right)^{-1}. \tag{4.37}$$

The resulting solution for the velocity field is

$$u(y) = R u^w \left(\frac{y}{h} + Q \frac{\ell}{h} \frac{\sinh(y/\ell)}{\cosh(h/\ell) + c_a \sinh(h/\ell)} \right). \tag{4.38}$$

Likewise, for α one finds

$$\alpha(y) = -\tau_{\text{ap}} R \frac{u^w}{h} \left(1 - \frac{\cosh(y/\ell)}{\cosh(h/\ell) + c_a \sinh(h/\ell)} \right). \tag{4.39}$$

4.2. APPARENT SLIP OF THE ISOTROPIC STATE SUBJECTED TO A FLOW

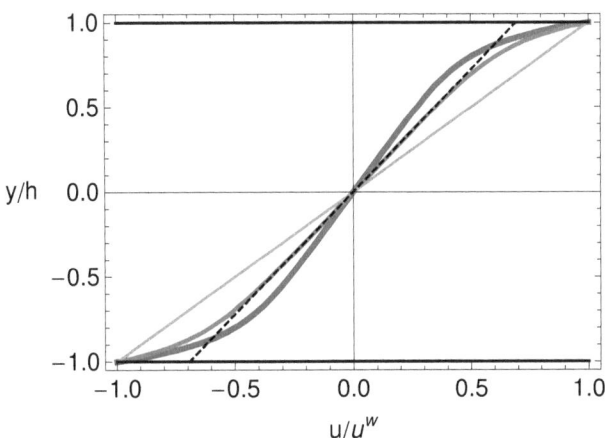

Figure 4.3: The velocity vs the distance in units of h is plotted for plane Couette flow, cf. (4.38). The model parameter are $Q = 8$, $h/\ell = 9$ and $c_\mathrm{a} = 0$ (thick blue line), $c_\mathrm{a} = 1$ (middle red line), $c_\mathrm{a} = 10$ (thin yellow line).

The velocity profile is plotted in Fig. 4.3 for $Q = 8$ and $h/\ell = 9$, corresponding to $2h/\ell_\mathrm{a} = 6$. The velocity is presented in units of the wall velocity u^w, the y-coordinate is in units of h, i.e. half the separation between the moving plates. The thick curve pertains to $c_\mathrm{a} = 0$, the other curves are for $c_\mathrm{a} = 1$, and $c_\mathrm{a} = 10$ (thin curve). Linear extrapolation of the velocity in the center towards the wall at $y = h$, e.g. see the dashed line shown for $c_\mathrm{a} = 0$, yields the velocity $h\gamma_0$ which is smaller than the wall velocity u^w. Thus one has an effective slip velocity

$$\delta u^\mathrm{eff} = u^\mathrm{w} - h\gamma_0 = (1-R)u^\mathrm{w} = (1-R)\gamma^\mathrm{ext}. \tag{4.40}$$

An effective slip length ℓ_u^eff defined by

$$\delta u^\mathrm{eff} = \ell_u^\mathrm{eff}\, \gamma^\mathrm{ext}. \tag{4.41}$$

From the relations above follows

$$\ell_u^\mathrm{eff} = \ell\, Q\, \tanh(h/\ell) \left(1 + (c_\mathrm{a} + Q\frac{\ell}{h})\tanh(h/\ell)\right)^{-1}. \tag{4.42}$$

For $\ell \ll h$, this expression reduces to

$$\ell_u^\mathrm{eff} \to \frac{\ell\, Q}{1 + c_\mathrm{a} + Q\frac{\ell}{h}}. \tag{4.43}$$

Clearly, the apparent slip is largest for $c_\mathrm{a} = 0$. For $c_\mathrm{a} \gg 1$, on the other hand, the simple Couette flow profile is approached as Fig. 4.4 implies.

CHAPTER 4. SPATIALLY INHOMOGENEOUS DYNAMICS OF NON-POLAR HARD-ROD FLUIDS

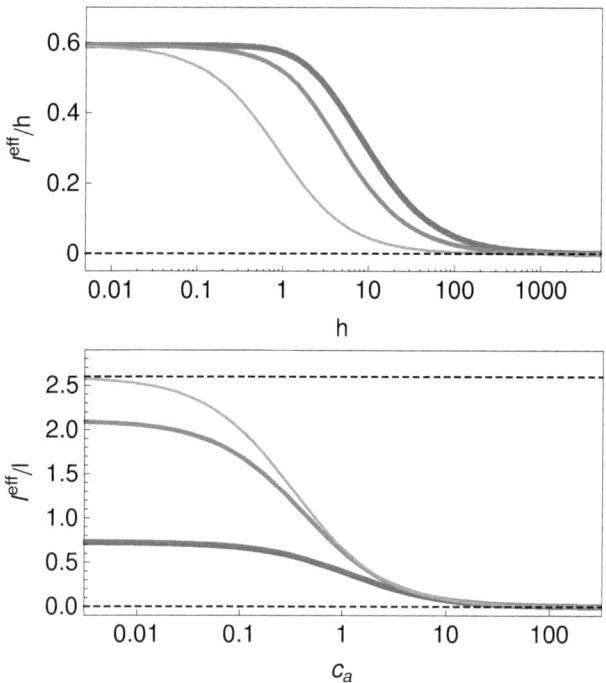

Figure 4.4: Upper: The effective slip length (4.42) in units of h vs h is displayed for plane Couette flow. The model parameter are $Q = 8$ and $c_a = 0$ (thick blue line), $c_a = 1$ (middle red line), $c_a = 10$ (thin yellow line). Lower: The effective slip length in units of ℓ is plotted as a function of the parameter c_a for the same conditions as in Fig. 4.3. The parameters are $Q = 8$, $\ell = 2$ and $h = 3$ (thick blue line), $h = 30$ (middle red line), $h = 300$ (thin yellow line).

4.2. APPARENT SLIP OF THE ISOTROPIC STATE SUBJECTED TO A FLOW

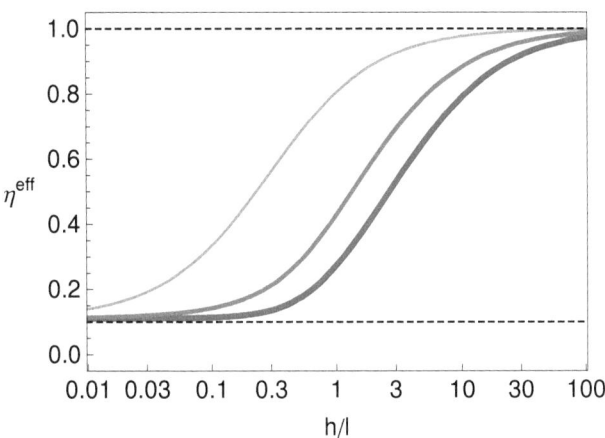

Figure 4.5: The reduced effective viscosity η_{eff}/η as a function of h/l for plane Couette flow is displayed. The parameter are $Q = 2$, and $c_a = 0$ (thick blue line), $c_a = 1$ (middle red line), $c_a = 10$ (thin yellow line).

In Fig. 4.4, the effective slip length in units of h as a function of the Couette cell size h is displayed. For small ratio ℓ_{eff}/h, the effective slip length have a significant effect on the flow. For higher values of the ratio the effect become more and more marginal. The slip effect can be disregarded for high ratio of the systems length compared to the molecular length.

An effective viscosity can be defined via the yx-component of the pressure tensor divided by the external shear rate, viz.:

$$\eta_{\text{eff}} = \frac{-p_{yx}}{\gamma^{\text{ext}}} = p_{\text{kin}} \left(\tau_0 \gamma(h) - \frac{\tau_{\text{ap}}}{\tau_{\text{a}}} \alpha(h) \right) / \gamma^{\text{ext}}. \quad (4.44)$$

The solutions given above for $\gamma(y)$ and $\alpha(y)$ eventually lead to

$$\eta_{\text{eff}} = R\eta. \quad (4.45)$$

In the limit $\ell \ll h$ one has

$$\eta^{\text{eff}} \to \eta \frac{1 + c_a}{1 + c_a + Q\frac{\ell}{h}}. \quad (4.46)$$

For $c_a \gg 1$, where the simple Couette flow profile is recovered, the effective viscosity becomes equal to the shear viscosity.

In Fig. 4.5, the effective viscosity is plotted for $Q = 2$, $\eta = 1$ and several values of $c_a = 0, 1, 10$. It can be recognized that for high values of the ratio h/ℓ the effective viscosity approaches the shear viscosity η. But if the magnitude of h is comparable to ℓ, the effective viscosity is much smaller than the shear viscosity. This effect depends strongly on the parameter c_a. For high c_a, this effect is negligible whereas for small values of c_a the shear viscosity

CHAPTER 4. SPATIALLY INHOMOGENEOUS DYNAMICS OF NON-POLAR HARD-ROD FLUIDS

is effected dramatically by the influence of the boundary conditions, as soon as $Q \neq 0$. If the first an the second Non-Newtonian viscosity are equal and therefore $Q = 0$ (no coupling of the alignment on the velocity), the effective viscosity is not influenced by the boundary conditions as expected.

4.2.5 Plane Poiseuille Flow

Consider now a flow between (identical) flat plates located at $y = h$ and $y = -h$. In the Poiseuille case, the walls are at rest and the flow is driven by the constant pressure gradient $\delta P/L$, where $P = P(x)$ is assumed. Notice that one has $\delta P = P(L) - P(0) < 0$ for a flow in the x-direction. Taking the symmetry of the problem into account, the ansatz

$$\alpha(y) = \alpha_2 \frac{y}{h} + \alpha_3 \sinh(y/\ell) \tag{4.47}$$

is now made for $\alpha(y)$. Similarly, for the shear rate one writes

$$\gamma(y) = \gamma_2 \frac{y}{h} + \gamma_3 \sinh(y/\ell) . \tag{4.48}$$

With the no-slip condition $u(h) = u(-h) = 0$ taken into account, the resulting velocity field is given by

$$u(y) = \frac{1}{2}\gamma_2 h \left(\frac{y^2}{h^2} - 1\right) + \gamma_3 \ell \left(\cosh(y/\ell) - \cosh(h/\ell)\right) . \tag{4.49}$$

The coefficients $\alpha_2, \alpha_3, \gamma_2, \gamma_3$ have to be determined from the differential equations (4.26), (4.27) and the boundary condition (4.28) with (4.29). In particular, the solution of the inhomogeneous equation (4.27) leads to

$$\alpha_2 = -\tau_{\text{ap}} h \eta^{-1} \frac{\delta P}{L} , \tag{4.50}$$

the homogeneous part of this differential equation is already obeyed by the ansatz (4.47). The boundary condition (4.28) leads to

$$\alpha_2 = -\alpha_3 \left(\sinh(h/\ell) + c_{\text{a}} \cosh(h/\ell)\right) . \tag{4.51}$$

From the differential equation (4.34) follows

$$\alpha_2 = -\tau_{\text{ap}} \gamma_2, \quad Q \alpha_3 = \tau_{\text{ap}} \gamma_3 \tag{4.52}$$

Insertion of these relations into (4.51) yields

$$\gamma_2 = h \eta^{-1} \frac{\delta P}{L}, \quad Q \gamma_2 = \gamma_3 \left(\sinh(h/\ell) + c_{\text{a}} \cosh(h/\ell)\right) . \tag{4.53}$$

The resulting solution for the velocity field is

$$u(y) = -\eta^{-1} \frac{\delta P}{L} h^2 \left(\frac{1}{2}(1 - \frac{y^2}{h^2}) + Q \frac{\ell}{h} \frac{\cosh(h/\ell) - \cosh(y/\ell)}{\sinh(h/\ell) + c_{\text{a}} \cosh(h/\ell)}\right) . \tag{4.54}$$

4.2. APPARENT SLIP OF THE ISOTROPIC STATE SUBJECTED TO A FLOW

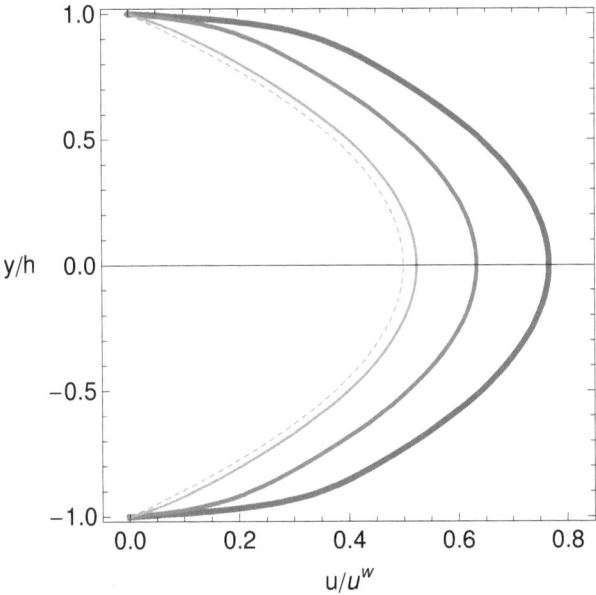

Figure 4.6: The Poiseuille flow profile (4.54) is displayed for several values of the parameter c_a: $c_\mathrm{a} = 0$ (thick blue line), $c_\mathrm{a} = 1$ (middle red line), $c_\mathrm{a} = 10$ (thin yellow line) and $c_\mathrm{a} = \infty$ (thin dashed line). The remaining parameters are $Q = 8$, $h/\ell = 9$.

CHAPTER 4. SPATIALLY INHOMOGENEOUS DYNAMICS OF NON-POLAR HARD-ROD FLUIDS

Clearly, for $Q > 0$ the flow is faster than for $Q = 0$. In Fig. 4.6 the Poiseuille flow profile is plotted for $Q = 8$ and different values of the boundary parameter c_a. For $c_a = 0$, the slip length is high and the flow is much faster than the ordinary Poiseuille flow without slip. The flow slows down for higher values of c_a and reaches the Poiseuille flow without slip in the limit $c_a \to \infty$.

Let $V(Q)$ be the flow velocity in the middle, i.e. for $y = 0$ as function of the model parameter Q. From (4.54) one infers for the effective slip velocity

$$\delta u^{\text{eff}} = V(Q) - V(0) = V(0)\, 2Q \frac{\ell}{h} \frac{\cosh(h/\ell) - 1}{\sinh(h/\ell) + c_a \cosh(h/\ell)}\,, \quad V(0) = -\eta^{-1}\frac{\delta P}{2L}. \quad (4.55)$$

The solution for α is

$$\alpha(y) = -\tau_{\text{ap}}\, h\, \eta^{-1} \frac{\delta P}{L}\left(\frac{y}{h} - \frac{\sinh(y/\ell)}{\sinh(h/\ell) + c_a \cosh(h/\ell)} \right). \quad (4.56)$$

The flux $J = J(Q)$ per unit length in the transverse direction is given by

$$J = \int_{-h}^{h} dy\, u(y) = 2\int_{0}^{h} dy\, u(y) = -\frac{2}{3}h^{3}\eta^{-1}\frac{\delta P}{L}\left(1 + 3Q\frac{\ell}{h}\frac{1 - \frac{\ell}{h}\tanh(h/\ell)}{c_a + \tanh(h/\ell)}\right). \quad (4.57)$$

When the velocity is not coupled with the alignment ($Q = 0$) but when a velocity slip characterized by the slip length ℓ_u is considered, the ratio of the flux with and without slip is given by $1 + 3\ell_u/h$. The relation (4.57) corresponds to such an expression, now with an effective slip length

$$\ell_u^{\text{eff}} = \ell Q \frac{1 - \frac{\ell}{h}\tanh(h/\ell)}{c_a + \tanh(h/\ell)}. \quad (4.58)$$

For $\ell \ll h$, this expression reduces to

$$\ell_u^{\text{eff}} \to \frac{\ell Q}{1 + c_a}. \quad (4.59)$$

In Fig. 4.7, the dependence of the effective slip length on h is displayed. As in previous plots the influence is most significant for $c_a = 0$ and less for higher values. The effective slip length saturates for relatively small values of the ratio h/ℓ to a value given by Eq. 4.59. This implies that the ratio of ℓ^{eff}/h goes to zero as in the Couette flow and for high values of h the apparent slip could be neglected.

Since $J \sim \eta^{-1}$, an effective viscosity for the Poiseuille flow is defined by

$$\eta_{\text{eff}} = (J(0)/J(Q))\, \eta = \eta \left(1 + 3Q\frac{\ell}{h}\frac{1 - \frac{\ell}{h}\tanh(h/\ell)}{c_a + \tanh(h/\ell)}\right)^{-1}. \quad (4.60)$$

In the limit $\ell \ll h$ one has

$$\eta_{\text{eff}} \to \eta \frac{1 + c_a}{1 + c_a + 3Q\frac{\ell}{h}}. \quad (4.61)$$

4.2. APPARENT SLIP OF THE ISOTROPIC STATE SUBJECTED TO A FLOW

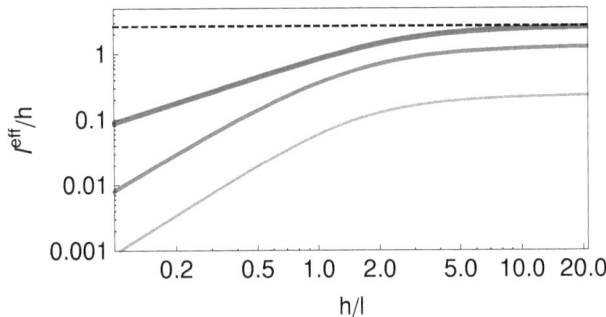

Figure 4.7: The effective slip length (4.58) vs h is plotted. The same conditions as in Fig. 4.6 are considered with parameters $Q = 8$ and $c_a = 0$ (thick blue line), $c_a = 1$ (middle red line), $c_a = 10$ (thin yellow line).

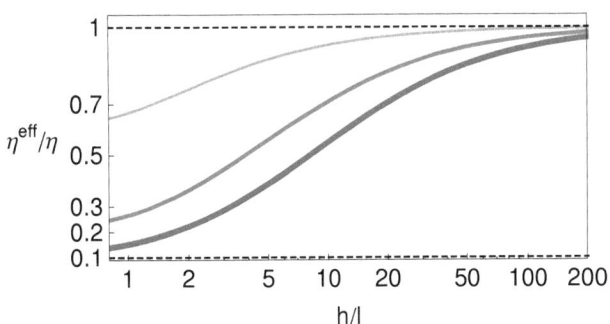

Figure 4.8: The reduced effective viscosity η^{eff}/η (4.60) as a function of h/ℓ is plotted. The parameters are $Q = 8$ and $c_a = 0$ (thick blue line), $c_a = 1$ (middle red line), $c_a = 10$ (thin yellow line).

CHAPTER 4. SPATIALLY INHOMOGENEOUS DYNAMICS OF NON-POLAR HARD-ROD FLUIDS

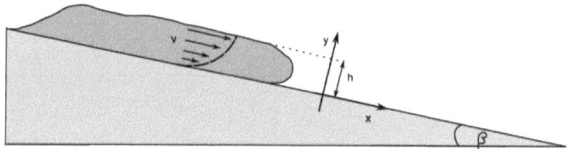

Figure 4.9: Sketch of the flow down an inclined plane.

In Fig. 4.8 the effective viscosity vs. h/ℓ is plotted. Again, for high values of h compared to ℓ a region is arrived where the apparent slip effect is not important but for smaller values of h/ℓ the effective viscosity decreases. The lowest value of the effective viscosity is $\eta^{\text{eff}} = \eta/9$, which is attained for $c_a = 0$. For $c_a \neq 0$, the effective viscosity reaches a minimum and increase after that minimum for smaller values of h/ℓ again. This behavior can be observed for very small values of ratios h/ℓ which are not physically relevant here and might be an artefact of the approximations employed. The reason is that in the limit $h \to 0$ the value η^{eff} depends on the parameter c_a. The limit value is $1/9$ for $c_a = 0$ whereas the limit value is 1 if $c_a \neq 0$.

4.2.6 Flow Down an Inclined Plane

For Newtonian fluids, the analysis of the gravity-driven flow down an inclined plane can be done in a manner similar to that of the Poiseuille flow. Here, however, different boundary conditions at the free surface lead to different flow profiles compared to the Poiseuille flow.

The x and y-axis parallel and perpendicular to the inclined plane is chosen, respectively, such that $y = 0$ defines the solid plane, while the free surface of the fluid film is located at $y = h$. The fluid film is assumed to be thin enough such that variations of the gravity force density $\mathbf{F} = \rho g(\sin\beta, -\cos\beta, 0)$ within the fluid layer can be neglected. The angle that the inclined plane forms with the horizontal axis is denoted by β. An illustration is given in Fig.(4.9).

Assuming incompressibility and no mechanical slip at the bottom plane, the velocity field is again of the form $\mathbf{v} = u(y)\mathbf{e}^x$.

At the free surface $y = h$, the following boundary conditions are imposed: First, the scalar pressure p must match the atmospheric pressure p_0. Second, it is required that no tangential stresses exist at the free surface, $P_{yx}|_h = 0$. Third, the boundary condition (4.28) for the alignment flux at the free surface $a(h) = -c_h \ell a'(h)$ is different from that at the bottom plane $a(0) = c_0 \ell a'(0)$ due to the different interactions the fluid experiences near the confining wall (c_0) and near the free surface (c_h).

From the y-component of the momentum balance equation, the profile of the scalar pressure $p(y) = p_0 + \rho g(h-y)\cos\beta$ is found to be the same as for a Newtonian liquid.

Inserting the constitutive relations (1.47) and (1.48), the second boundary condition at

4.2. APPARENT SLIP OF THE ISOTROPIC STATE SUBJECTED TO A FLOW

the free surface becomes $-2\eta_{\text{iso}}\Gamma_{yx}|_h + \sqrt{2}p_{\text{kin}}(\tau_{\text{ap}}/\tau_{\text{a}})\Phi_{yx}|_h = 0$. Note, the velocity gradient need not vanish at the free surface due to the alignment contribution to the pressure tensor.

In the present case, the stationary solution to (4.16) reduces not to (4.26) but $\alpha - \ell\alpha'' = G(h-y)$, with $G = \rho g \tau_{\text{ap}} \sin\beta/\eta$. The general solution to this equation reads $\alpha(y) = \alpha_1 e^{y/\ell} + \alpha_2 e^{-y/\ell} + G(y-h)$. The boundary conditions lead to the final expressions

$$\alpha(y) = \frac{G\ell}{Z}\left[-c_h \sinh(y/\ell) - c_0 c_h \cosh(y/\ell) - (c_0 + \frac{h}{\ell})\sinh([y-h]/\ell)\right.$$
$$\left. + c_h(c_0 + \frac{h}{\ell})\cosh([y-h]/\ell)\right] + G(y-h) \quad (4.62)$$

with $Z = (1 + c_0 c_h)\sinh(h/\ell) + (c_0 + c_h)\cosh(h/\ell)$.

The resulting solution for the velocity field is

$$u(y) = u_0\left[\frac{1}{2}(2 - \frac{y}{h})\frac{y}{h} + Q\left(\frac{\ell}{h}\right)^2 \frac{1}{Z}F(y/\ell)\right] \quad (4.63)$$

where $u_0 = \rho g h^2 \sin\beta/\eta$ is twice the velocity at the free surface for Newtonian fluids, $u(y = h, Q = 0) = u_0/2$. The modification of the flow profile due to the alignment is described by

$$F(y/\ell) = c_h[1 - \cosh(y/\ell)] - c_0 c_h \sinh(y/\ell) + (c_0 + \frac{h}{\ell})\{\cosh(h/\ell) - \cosh([y-h]/\ell)\}$$
$$+ c_h(c_0 + \frac{h}{\ell})\{\sinh(h/\ell) + \sinh([y-h]/\ell)\}. \quad (4.64)$$

The flux $J(Q)$ down the inclined plane per unit length in the transverse direction is found to be given by

$$J(Q) = \int_0^h dy\, u(y) = \frac{1}{3}u_0 h\left[1 + \frac{3Q}{Z}\frac{\ell}{h}\{k_1 \sinh(h/\ell) + k_2 \cosh(h/\ell)\}\right] \quad (4.65)$$

with $k_1 = c_h(1 + c_0\ell/h) - (\ell/h)k_2$, $k_2 = 1 + (c_0 + c_h)\ell/h$. From the flux J, the apparent slip velocity is defined by [154]

$$\delta u^{\text{eff}} = \left.\frac{\partial(J/h^2)}{\partial(1/h)}\right|_{\rho g h \sin\beta} \quad (4.66)$$

where the derivative is taken at constant surface shear stress (in absence of coupling). From Eq. (4.65) one finds

$$\delta u^{\text{eff}} = u_0 \frac{Q}{Z}\left[k_1 k_2 + (c_h^2 + k_2^2)(\ell/h)^2 \cosh(h/\ell)\sinh(h/\ell)\right]. \quad (4.67)$$

A particularly interesting case is a vanishing alignment flux at the bottom plane and vanishing alignment flux gradient at the surface. In this case, the boundary conditions become $\alpha(0) = 0$, $\alpha'(h) = 0$, then one has

$$u^\infty(y) = \frac{\rho g \sin\beta}{\eta}h^2\left(\frac{1}{2}(2 - \frac{y}{h})\frac{y}{h} + Q\frac{\ell^2}{h^2}\left\{\frac{h}{\ell}\sinh(y/\ell) - \cosh(y/\ell) + 1\right\}\right). \quad (4.68)$$

CHAPTER 4. SPATIALLY INHOMOGENEOUS DYNAMICS OF NON-POLAR HARD-ROD FLUIDS

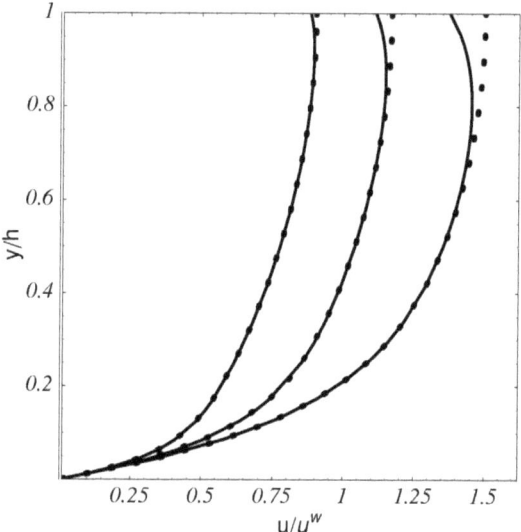

Figure 4.10: The flow profile of the flow down an inclined plane is displayed for several values of the parameter $h/\ell = 8, 12, 20$ (from the right to the left). The remaining parameters are $c_0 = c_a = 0$, $c_h = \infty$, $Q = 8$. The dashed line shows the half of the Poiseuille flow profile for the same parameters.

Note, that the special case (4.68) can be derived from Eq. (4.63) for $c_0 = 0$, $c_h \to \infty$ only in the limiting case $h/\ell \ll 1$. In this case, the apparent velocity slip (4.67) simplifies to

$$\delta u^{\infty,\text{eff}} = u_0 Q \left(\frac{\ell}{h}\right)^2 \left[(2 - \frac{\ell}{h})\cosh(h/\ell) + (1 - 3\frac{\ell}{h})\sinh(h/\ell)\right]. \tag{4.69}$$

In this approximation the flow profile is very similar to the plane Poiseuille flow cut in the middle plane and for $h/\ell < 0.1$ it cannot be distinguished in graphical diagrams. In the limiting case $h/\ell \gg 1$, $c_0 = 0$ and $c_h \to \infty$ the influence of boundary conditions become smaller and for $h/\ell > 1000$ the flow profile coincides with the plane Poiseuille flow cut in the middle for graphical accuracy. In Fig. 4.10 the flow profile for the flow down an inclined plane is compared with Poiseuille flow cut in the middle (dashed line) between the limiting cases for $c_0 = 0$ and $c_h \to \infty$. The difference between the Poiseuille flow and the inclined flow profile is significant and becomes smaller for higher values of the ratio h/ℓ.

4.2. APPARENT SLIP OF THE ISOTROPIC STATE SUBJECTED TO A FLOW

4.2.7 Alignment

For the special cases considered in this section, viz.: isotropic phase and small shear rates, only the xy-component of the alignment tensor is affected by the flow. The alignment tensor is written as, cf. (4.25),

$$a_{\mu\nu} = \sqrt{2}\, a(y)\, \overline{e_\nu^x e_\mu^y} + \dots \tag{4.70}$$

The ellipses stand for the other components which could be non-zero, due the influence of the walls. In the isotropic phase, the relation

$$\Phi_{\mu\nu}(\mathbf{a}) = A a_{\mu\nu} + \dots - \xi_a^2 \triangle a_{\mu\nu} \tag{4.71}$$

is equivalent to

$$a - \xi^2 a'' = A^{-1} \alpha\,, \quad \xi^2 = A^{-1} \xi_a^2\,, \tag{4.72}$$

where the elastic coherence length ξ is used. For $\xi = 0$ the alignment $a(y)$ is determined by the quantity $\alpha(y)$ as computed above. For $\xi \neq 0$, on the other hand, the value of a, i.e. of the xy-component of the alignment tensor can and has to be prescribed in order to obtain a unique solution for the alignment. It is recalled that the boundary condition given above was not formulated directly for the alignment tensor but rather for the derivative of the potential function with respect to the alignment.

As before, the plane Couette flow and plane Poiseuille flow geometries are treated. In analogy to (4.30) the ansatz

$$a(y) = a_0 + a_1 \cosh(y/\ell) + b \cosh(y/\xi)\,, \quad \xi \neq \ell\,, \tag{4.73}$$

or

$$a(y) = a_0 + a_1\, (y/\ell)\, \sinh(y/\ell) + b \cosh(y/\ell)\,, \quad \xi = \ell \tag{4.74}$$

is made for the Couette case. Again it is assumed that the walls at $y = h$ and at $y = -h$ are identical. The coefficients a_0, a_1 follow from the differential equation (4.72), the coefficient b is fixed by the value a^w of the alignment at the wall. More specifically, one obtains

$$a_0 = A^{-1}\alpha_0\,, \quad a_1 = \left(1 - (\xi^2/\ell^2)\right)^{-1} A^{-1} \alpha_1\,, \quad \xi \neq \ell\,, \tag{4.75}$$

and

$$a_0 = A^{-1}\alpha_0\,, \quad a_1 = -\frac{1}{2} A^{-1} \alpha_1\,, \quad \xi = \ell\,, \tag{4.76}$$

Similarly, for the Poiseuille flow between identical walls it is used, cf. (4.47),

$$a(y) = a_2\, (y/\ell) + a_3 \sinh(y/\ell) + b_1 \sinh(y/\xi) + b_2 \cosh(y/\xi)\,, \quad \xi \neq \ell\,, \tag{4.77}$$

or

$$a(y) = a_2\, (y/\ell) + a_3\, (y/\ell) \cosh(y/\ell) + b_1 \sinh(y/\xi) + b_2 \cosh(y/\xi)\,, \quad \xi = \ell\,. \tag{4.78}$$

The differential equation (4.72) yields relations for the coefficients a_2, a_3 which are identical to (4.75) and (4.76) with the subscripts $0, 1$ replaced by $2, 3$. The coefficients b_1, b_2 are determined by the alignment at the wall. In Fig. 4.11 the xy-component of the alignment tensor for the Couette flow and the Poiseuille flow is displayed, respectively. At the boundary

CHAPTER 4. SPATIALLY INHOMOGENEOUS DYNAMICS OF NON-POLAR HARD-ROD FLUIDS

uniaxial alignment is used $\mathbf{a} = \sqrt{3/2}a_{\text{eq}}\overline{\mathbf{nn}}$. In our case the director $\mathbf{n}$ lies perpendicular to the xy-plane. The xy-component of the alignment tensor is then $a = 0$. The Fig. 4.11 show a significant effect on the flow alignment by the apparent slip parameter c_a. In the Couette flow the shape of the curve is only mildly affected whereas the minimum grow. This means for small values of c_a that the flow alignment angle of the molecules in the middle of the Couette cell deviates less from the flow alignment angle of the molecules at the wall. In the Poiseuille case, the alignment solutions are antisymmetric with respect to the middle plane as a consequence of the symmetry of the velocity profile. As in the Couette flow, curves for several parameter values c_a are between the limiting curves for $c_a \to \infty$ indicated by the dashed line and for $c_a = 0$ (thick line). High values of the parameter c_a lead to small flow alignment angle in the bulk except for $y = 0$ where all curves are intersecting.

4.2.8 Cylindrical Couette Flow Geometry

In order to determine material functions of Non-Newtonian fluids, it is common practice to carry out shear experiments in different geometries (cone-plate or cylindrical geometry). Here a fluid is considered between two coaxial cylinders which are in relative rotation (see, Fig. 4.12). The inner-cylinder radius is denoted by $r_i > 0$ and the outer-cylinder by r_o ($r_o > r_i$).

It is assumed that the cylinders are infinitely long to avoid boundary effects from the top and the bottom of the cylinder. According to the special symmetry polar coordinates are chosen. The appropriate velocity field is taken as

$$\mathbf{v} = (0, u(r), 0)^t \tag{4.79}$$

and the orthonormal tensor basis

$$\begin{aligned}\mathbf{T}^0 &= \sqrt{3/2}\,\overline{\mathbf{e}_z\mathbf{e}_z}, \quad \mathbf{T}^1 = 1/\sqrt{2}\,(\mathbf{e}_r\mathbf{e}_r - \mathbf{e}_\varphi\mathbf{e}_\varphi) \\ \mathbf{T}^2 &= \sqrt{2}\,\overline{\mathbf{e}_r\mathbf{e}_\varphi}, \quad \mathbf{T}^3 = \sqrt{2}\,\overline{\mathbf{e}_r\mathbf{e}_z}, \quad \mathbf{T}^4 = \sqrt{2}\,\overline{\mathbf{e}_\varphi\mathbf{e}_z},\end{aligned} \tag{4.80}$$

which is similar to the Cartesian orthonormal tensor basis used in 1.32. In analogy to the previous chapter a special Ansatz for the strain tensor $\boldsymbol{\Gamma}$ and for $\boldsymbol{\Phi}$ was made:

$$\boldsymbol{\Gamma} = \frac{1}{\sqrt{2}}\gamma(r)\mathbf{T}^2, \tag{4.81}$$

$$\boldsymbol{\Phi} = \alpha(r)\mathbf{T}^2. \tag{4.82}$$

With the help of this ansatz, Eqs. (4.14) and (4.16) can be rewritten as

$$\alpha(r)\mathbf{T}^2 - \ell_a^2 \mathbf{T}^2 \triangle_{r,\varphi}\alpha(r) - \ell_a^2 \alpha(r)\triangle_{r,\varphi}\mathbf{T}^2 = -\sqrt{2}\tau_{\text{ap}}\gamma(r)\mathbf{T}^2 \tag{4.83}$$

$$p_{\text{kin}}\sqrt{2}\frac{\tau_{\text{ap}}}{\tau_a}\nabla_{r,\varphi}\alpha(r)\cdot\mathbf{T}^2 + p_{\text{kin}}\sqrt{2}\frac{\tau_{\text{ap}}}{\tau_a}\alpha(r)\nabla_{r,\varphi}\cdot\mathbf{T}^2 = \eta_{\text{iso}}\triangle_{r,\varphi}\mathbf{v}, \tag{4.84}$$

where $\triangle_{r,\varphi}$ and $\nabla_{r,\varphi}$ are the Laplacian and the gradient in polar coordinates, respectively. Eq. (4.84) can be rearranged as

$$\nabla_{r,\varphi}\cdot\left(\eta_{\text{iso}}\boldsymbol{\Gamma} - 1/\sqrt{2}\frac{\tau_{\text{ap}}}{\tau_a}p_{\text{kin}}\alpha(r)\mathbf{T}^2\right) = 0, \tag{4.85}$$

4.2. APPARENT SLIP OF THE ISOTROPIC STATE SUBJECTED TO A FLOW

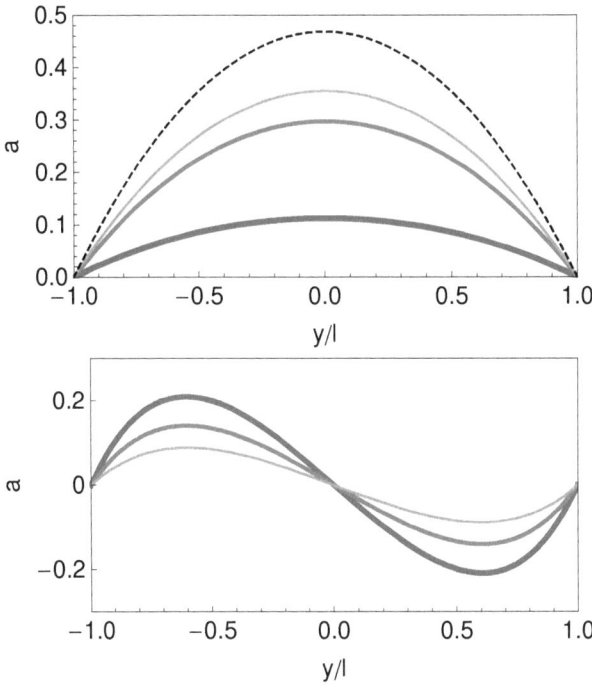

Figure 4.11: Upper panel: The xy-component of the alignment tensor is plotted for plane Couette flow for several values of the parameter c_a. The parameters are chosen as $Q = 8$, $A = 1$, $\xi = 0.3$, $\ell = 0.2$, $\tau_\mathrm{ap} = -0.5$, $u^\mathrm{w} = 1$, $h/\ell = 1$, $a(h) = 0$ and $c_\mathrm{a} = 0$ (thick blue line), $c_\mathrm{a} = 10$ (middle red line), $c_\mathrm{a} = 20$ (thin yellow line), $c_\mathrm{a} = \infty$ (dashed line). Lower panel: The xy-component of the alignment tensor is plotted for plane Poiseuille flow for several values of the parameter c_a. The parameter are $Q = 8$, $A = 1$, $\xi = 0.1$, $l = 0.2$, $\tau_\mathrm{ap} = 0.5$, $u^\mathrm{w} = 1$, $h = 1$, $a(h) = 0$ and $c_\mathrm{a} = 0$ (thick blue line), $c_\mathrm{a} = 1$ (middle red line), $c_\mathrm{a} = \infty$ (thin yellow line).

CHAPTER 4. SPATIALLY INHOMOGENEOUS DYNAMICS OF NON-POLAR HARD-ROD FLUIDS

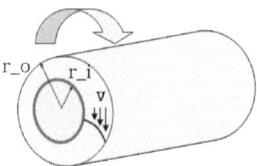

Figure 4.12: The cylindrical Couette flow geometry is given by to infinitely long coaxial cylinders which are in relative rotation. The radius r_i denotes the inner and the radius r_o the outer cylinder, respectively.

when the incompressibility condition is used. The solution of Eq. (4.85) is given by

$$\gamma(r) = \frac{1}{\eta_{iso}} \frac{\tau_{ap}}{\tau_a} p_{kin}\alpha(r) + c/r^2 \tag{4.86}$$

and leads with Eq. (4.83) to a Bessel differential equation for the function α,

$$\alpha(r) - \ell\triangle_{r,\phi}\alpha(r) = c/r^2 \tag{4.87}$$

or explicitly

$$r^2\alpha''(r) + r\alpha'(r) - (1 + \frac{r^2}{\ell^2})\alpha(r) = f(r). \tag{4.88}$$

In our case $f(r) = c/r^2$. The homogeneous solution are modified Bessel-functions (Bessel-functions with pure imaginary argument) which are denoted as $\mathfrak{J}_\lambda$ (first kind of order λ) and $\mathfrak{K}_\lambda$ (second kind of order λ). The particular solution of the Bessel-Equation can be derived by

$$\alpha^p(r) = \frac{\pi}{2}\mathfrak{K}_\lambda(r)\int r\mathfrak{J}_\lambda(r)f(r)dr - \frac{\pi}{2}\mathfrak{J}_\lambda(r)\int r\mathfrak{K}_\lambda(r)f(r)dr, \tag{4.89}$$

such that the general solution can be written as

$$\begin{aligned}\alpha(r) &= \alpha_1\mathfrak{J}_1(\hat{r}) + \alpha_1\mathfrak{K}_1(\hat{r}) \\ &+ \frac{\pi}{2}c\ell^2(\mathfrak{K}_1(\hat{r})\int \mathfrak{J}_1(\hat{r})/\hat{r}d\hat{r} - \mathfrak{J}_1(\hat{r})\int \mathfrak{K}_1(\hat{r})/\hat{r}d\hat{r}).\end{aligned} \tag{4.90}$$

Here the abbreviation $\hat{r} = r/\ell$ is used. The solution of the integrals are generalized hypergeometric functions and Maijer G-functions. In principle, it is possible to use the exact solution of α for further calculations. In many applications, however, cylindrical Couette flow is considered with large radii of r_i and r_o, i.e. $r_i, r_o \gg 1$. In that case, the modified Bessel-functions can be approximated by $\mathfrak{J}_\lambda(r) \sim \frac{1}{\sqrt{r}}e^r$ and $\mathfrak{K}_\lambda(r) \sim \frac{1}{\sqrt{r}}e^{-r}$ [155, 156], so that the integration is trivial and the general solution for not too small values of r_i, r_o is

$$\alpha(r) = \frac{1}{\sqrt{\hat{r}}}(\alpha_1 e^{\hat{r}} + \alpha_2 e^{-\hat{r}}) + \alpha_0. \tag{4.91}$$

4.2. APPARENT SLIP OF THE ISOTROPIC STATE SUBJECTED TO A FLOW

The integration constant was denoted by α_0. The coefficients α_1 and α_2 are determined by the boundary conditions for the alignment flow Φ

$$\alpha(r_\mathrm{i}) = c_\mathrm{a}\ell\alpha'(r_\mathrm{i}) \quad \alpha(r_\mathrm{o}) = -c_\mathrm{a}\ell\alpha'(r_\mathrm{o}). \tag{4.92}$$

The shear rate $\gamma(r)$ according to Eq.(4.86) is given by

$$\gamma(r) = -Q\frac{1}{\sqrt{\hat{r}}}(\alpha_1 e^{\hat{r}} + \alpha_2 e^{-\hat{r}}) + \gamma_1/r^2 + \gamma_0 \tag{4.93}$$

and the velocity profile can be derived by integration as

$$u(r) = -Q\sqrt{\pi\ell}(\alpha_1(c_\mathrm{a},\ell,r_\mathrm{i},r_\mathrm{o})\mathrm{erfi}[\sqrt{\hat{r}}] + \alpha_2(c_\mathrm{a},\ell,r_\mathrm{i},r_\mathrm{o})\mathrm{erf}[\sqrt{\hat{r}}]) + \gamma_0 r - \gamma_1/r, \tag{4.94}$$

where $\mathrm{erf}[x]$ denotes the error function and $\mathrm{erfi}[z] = -i\mathrm{erf}[iz]$. The coefficients γ_0 and γ_1 are determined by the no slip conditions of the velocity at the boundary r_i and r_o.

For a better comparison to the plane Couette flow considered above, new coordinates are introduced. The variables $r_\mathrm{o} - r_\mathrm{i} = 2h$, $\xi = \{r | r \in [r_\mathrm{i}, r_\mathrm{o}]\}$ and $r_m = r_\mathrm{i} + (r_\mathrm{o} - r_\mathrm{i})/2$ are denoted. The boundary conditions (4.92) reduce to

$$\alpha(r_\mathrm{m}, -h) = c_\mathrm{a}\ell\alpha'(r_\mathrm{m}, -h) \quad \alpha(r_\mathrm{m}, h) = -c_\mathrm{a}\ell\alpha'(r_\mathrm{m}, h). \tag{4.95}$$

It is assumed for the velocity boundary conditions $u(r_i = r_m - h) = u(r_\mathrm{m}, -h) = -u^\mathrm{w}$ and $u(r_o = r_m + h) = u(r_\mathrm{m}, h) = u^\mathrm{w}$ that is the outer cylinder is moved by u^w and the inner by $-u^\mathrm{w}$. In this co-ordinates the velocity profile is a function of ξ and with the parameters $c_\mathrm{a}, \ell, h, r_\mathrm{m}, u^\mathrm{w}$:

$$u(\xi; c_\mathrm{a}, \ell, h, r_\mathrm{m}, u^\mathrm{w}) = -Q\sqrt{\pi\ell}(\alpha_1(c_\mathrm{a},\ell,r_\mathrm{i},r_\mathrm{o})\mathrm{erfi}[\sqrt{\hat{\xi}+\hat{r}_m}] + \tag{4.96}$$
$$\alpha_2(c_\mathrm{a},\ell,r_\mathrm{i},r_\mathrm{o})\mathrm{erf}[\sqrt{\hat{\xi}+\hat{r}_m}]) + \gamma_0(r_\mathrm{m}+\xi) - \gamma_1/(r_\mathrm{m}+\xi).$$

CHAPTER 4. SPATIALLY INHOMOGENEOUS DYNAMICS OF NON-POLAR HARD-ROD FLUIDS

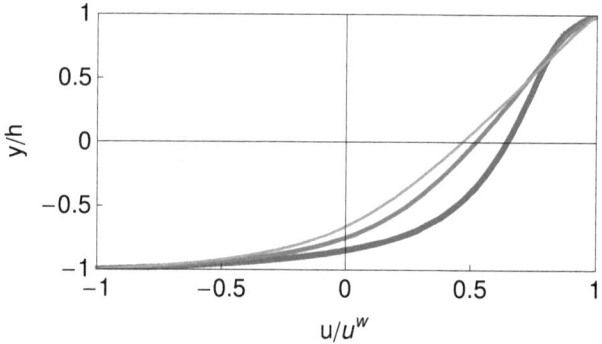

Figure 4.13: The velocity vs the distance in units of h is plotted for Couette flow in cylindrical geometry. The chosen model parameters are $Q = 8$, $h/\ell = 9$, $r_m/\ell = 10$ and $c_a = 0$ (thick blue line), $c_a = 3$ (middle red line), $c_a = 100$ (thin yellow line).

The coefficients α_1, α_2, γ_0 and γ_1 are analytically calculated with the computer algebra program "Mathematica6.0". In Fig. 4.13, the velocity profile is plotted for $Q = 8$, $h/\ell = 9$, $l = 1$ and $r_m/\ell = 10$. The velocity is presented in units of the wall velocity u^w and y denotes ξ/ℓ. The thick curve pertains to the highest possible boundary effect $(c_a = 0)$ for $Q = 8$ fixed. For high values of the coupling of the alignment to the pressure tensor the apparent slip becomes stronger as Fig. 4.14 shows. As in the plane Couette flow, the extrapolation of the velocity in the center towards the wall is smaller than the wall velocity u^w and for $c_a \to \infty$ the flow profile reaches the cylindrical Couette flow without apparent slip. In contradistinction to the plane Couette flow, the velocity profile in cylindrical geometry is asymmetric. That is a direct consequence of the radial geometry, i.e. of the term $1/(r_m + \xi)$. In the case where $r_m \ggg h$ it can be approximated $1/(r_m + \xi) \approx 1/r_m$ so that the results for the cylindrical Couette flow can be compared to the results of plane Couette flow directly.

In particular, the velocity profile cannot be distinguished from the plane Couette flow profile in Fig. 4.3 if the same parameter values are used.

4.3 Orientational Dynamics and Flow Properties of Nematic State

4.3.1 Imposed Shear

In the last section the boundary effect due to the alignment flux was studied for non-polar hard-rod fluids in the isotropic phase subjected to different flows. Here the orientational behavior of the nematic phase ($\vartheta = 0$) subjected to a shear flow is addressed. Firstly, it is assumed that the second Newtonian viscosity ν_{iso} is high enough to approximate the flow

4.3. ORIENTATIONAL DYNAMICS AND FLOW PROPERTIES OF NEMATIC STATE

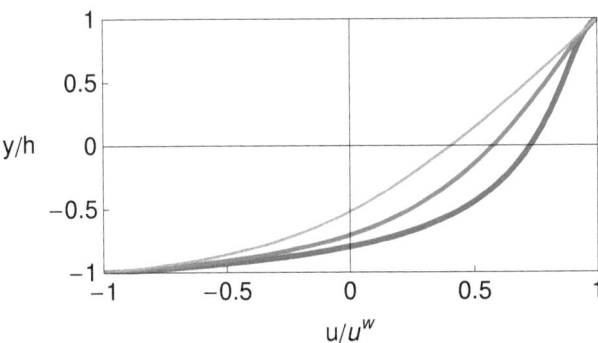

Figure 4.14: Same as Fig. 4.13 but for parameters $r_m/\ell = 11$, $h/\ell = 9$, $c_a = 0$ and $Q = 10$ (thick blue line), $Q = 1$ (middle red line), $Q = 0$ (thin yellow line).

profile by a linear Couette flow profile ($u = \dot{\gamma} y$), such that the shear rate is constant between the two plates.

Due to the nonlinear nematic potential one has no reasonable chance to find analytical solutions. The dynamical equations (1.84) for the alignment tensor are solved numerically by 4th order finite difference scheme and 4th order Runge-Kutter-Fehlberg integration (for a detailed explanation see the appendix). For the numerical solution initial and boundary conditions for all variables have to be specified.

In the simplest approach for the orientational boundary condition a specific equilibrium value is fixed at the plates, this is referred to as strong anchoring condition. For the boundary values it is distinguished between planar anchoring (uniaxial alignment within the plate plane, here into the x-direction), homeotropic anchoring (alignment normal to the plates), isotropic anchoring (modeling rough surfaces, zero alignment tensor) and degenerate anchoring (particle axes equally distributed in the plate plane) [157], i.e.

$$\text{planar anchoring}: \quad \mathbf{a}(-h,t) = \mathbf{a}(h,t) = \sqrt{3/2}a_{eq}\overline{\mathbf{e}_x \mathbf{e}_x} \quad (4.97)$$
$$\text{homeotropic anchoring}: \quad \mathbf{a}(-h,t) = \mathbf{a}(h,t) = \sqrt{3/2}a_{eq}\overline{\mathbf{e}_y \mathbf{e}_y} \quad (4.98)$$
$$\text{denenerate anchoring}: \quad \mathbf{a}(-h,t) = \mathbf{a}(h,t) = -\sqrt{3/2}a_{eq}\overline{\mathbf{e}_y \mathbf{e}_y} \quad (4.99)$$
$$\text{isotropic anchoring}: \quad \mathbf{a}(-h,t) = \mathbf{a}(h,t) = 0 \quad (4.100)$$

To be consistent with the boundary conditions, the boundary values of the variables are used as initial orientation with small perturbations at every space point and zero plate speed. After a relaxation time of 10 time units the flow is switched on by moving the plates. To simplify the analysis the parameter D_a is set to be zero (no effect of the alignment flux is considered).

Due to the boundaries the spatio-temporal solutions are strongly different from the characteristic bulk solutions. For a similar model and planar anchoring conditions Rey *et. al.*

CHAPTER 4. SPATIALLY INHOMOGENEOUS DYNAMICS OF NON-POLAR HARD-ROD FLUIDS

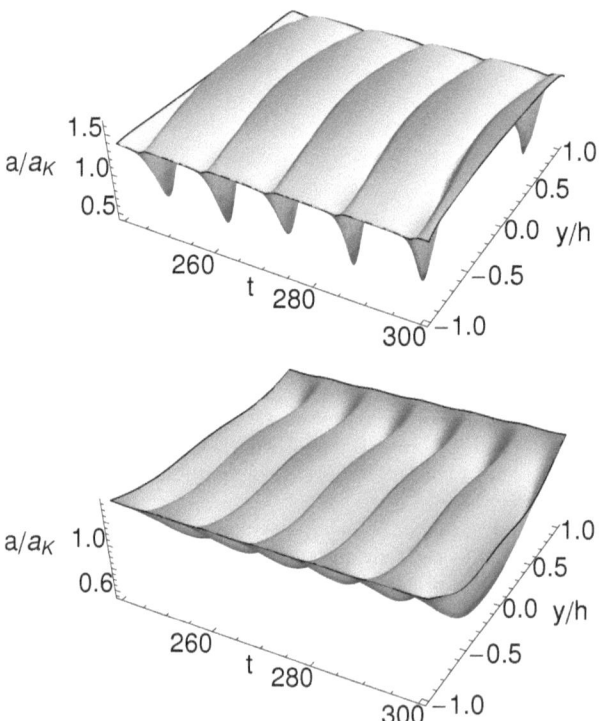

Figure 4.15: The spatio-temporal dynamics of the order parameter $a = \sqrt{\mathbf{a} : \mathbf{a}}$ for the tumbling parameter $\lambda_{\mathrm{K}} = 1$ and Ericksen number $Er = 100$ show in the middle region between the plates a tumbling and in the boundary region a wagging motion of the principal director (upper graph). The T/W-transition region is characterized by strongly decreasing values of the order parameter a. The Weissenberg number $Wi = 1$ and $\vartheta = 0$, $\kappa_{\mathrm{a}} = 0$. For higher shear rates $Wi = 5$ the tumbling motion disappears (lower graph).

4.3. ORIENTATIONAL DYNAMICS AND FLOW PROPERTIES OF NEMATIC STATE

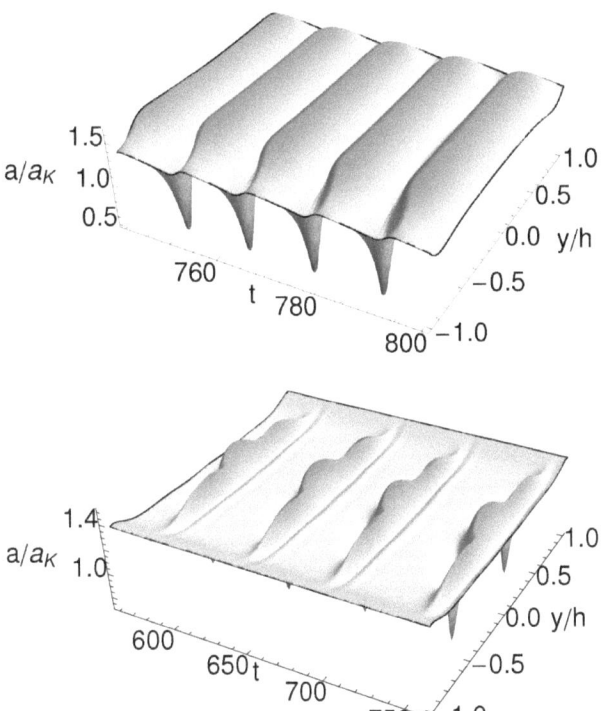

Figure 4.16: The spatio-temporal dynamics for the order parameter a for $Er = 1000$ is displayed. The upper figure show in the middle region kayaking-tumbling behavior for the model parameters $\lambda_K = 1.0$, $Wi = 2.0$, $\vartheta = 0$ and κ_a. The lower panel shows kayaking wagging orientational dynamics. The model parameters are $\lambda_K = 1.3$, $\kappa_a = 0$, $\vartheta = 0$ and $Wi = 3$.

CHAPTER 4. SPATIALLY INHOMOGENEOUS DYNAMICS OF NON-POLAR HARD-ROD FLUIDS

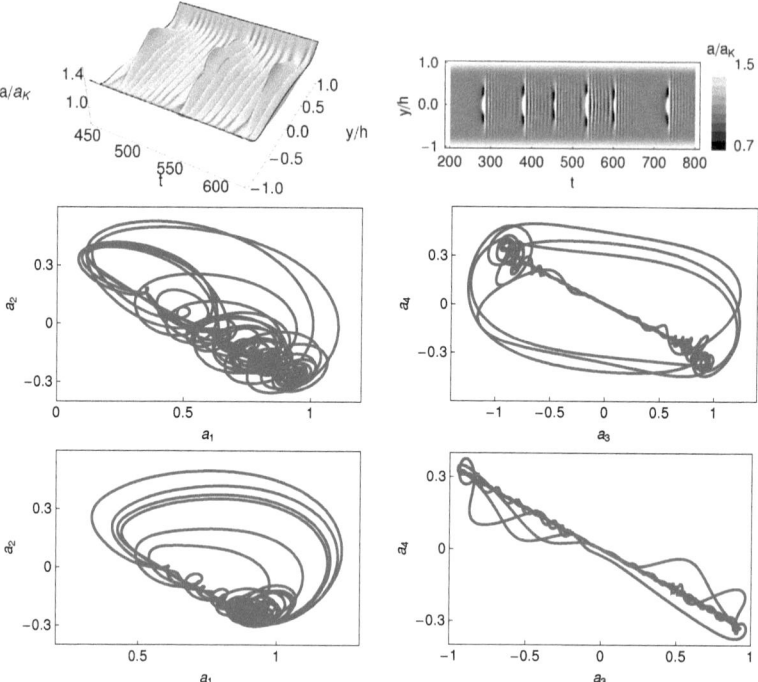

Figure 4.17: For the model parameters $\lambda_K = 1.175$, $\kappa_a = 0$, $Wi = 3.75$, $\vartheta = 0$ and $Er = 1000$ the orientational dynamics is irregular. The upper panel shows the time evolution of the oder parameter a and the lower the corresponding orbits of the alignment tensor components. The orbit for $y = 0$ is rather irregular whereas closer to the plates $y = 0.6$ the dynamics is restricted. The spatio-temporal dynamics for the order parameter a for $Er = 1000$ is displayed. The model parameters are $\lambda_K = 1.175$, $Wi = 3.75$.

4.3. ORIENTATIONAL DYNAMICS AND FLOW PROPERTIES OF NEMATIC STATE

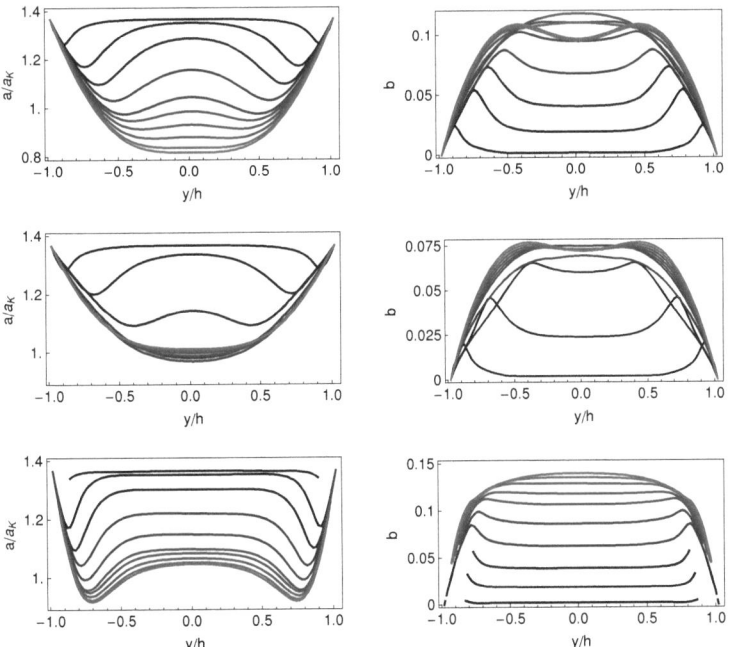

Figure 4.18: The figure shows the time averaged profile (over 150 time units) of the order parameter a (left) and the biaxiality parameter b (right) for different values of the tumbling parameter λ_K, the Weissenberg Wi number and Ericksen number Er. In every plot Wi is varied from black to blue via $Wi = 0.1, 1.1, ..., 10.1$. The first two panels from the top are for $\lambda_K = 0.8, 1.0$ and for $Er = 100$. The parameter values of the third (last) panel are determined by $\lambda_K = 0.8$ and $Er = 500$. The parameter $\kappa_a = 0$ for all figures.

CHAPTER 4. SPATIALLY INHOMOGENEOUS DYNAMICS OF NON-POLAR HARD-ROD FLUIDS

give a classification of characteristic solutions [41, 57–59], including in-plane and out of plane flow modes:

- In-plane elastic-driven steady state
- In-plane tumbling-wagging composite state
- In-plane viscous-driven steady state
- Out-of-plane elastic driven steady state with a-chiral structure
- Out-of-plane elastic-driven steady state with chiral structure
- Out-of-plane tumbling-wagging composite state with periodic chirality
- Out-of-plane tumbling-wagging composite state with π chiral structure

In addition, three in-plane flow modes are found and reported by Zang et. al. [158]. For the kinetic Doi-model similar characteristic solutions are identified [98, 159, 160]. In the present model most of the flow modes could be identified in the simulations (the chiral character was not considered). Beside the oscillating flow modes irregular and chaotic orientational behavior were found for specific parameter values. Recently, spatio-temporal chaotic solutions (rheochaos) have been observed numerically and in experiments [16, 24, 54, 56, 161]. In the following the focus is on the composite flow modes rather than the classification. As discussed later, composite flow modes show an interesting defect (planar biaxiality) structure. Defects have a deep impact on the flow properties of the fluid (for e.g. defect stress correlation, see [162]) and therefore are of particular interest.

In Fig. 4.15 the spatio-temporal dynamics of the order parameter for in-plane tumbling-wagging composite flow mode and in-plane tumbling flow mode is shown. In the upper figure the principal director rotates (tumbling) in the middle region of the pore and oscillates (wagging) in the region close to the plates. The transition from tumbling to wagging is characterized by very low values of the order parameter $a = \sqrt{\mathbf{a} : \mathbf{a}}$ and a high biaxiality. The small Ericksen number $Er = 100$ (small plate separation compared to the elasticity length scale) gives rise to in-plane motion of the principal director although the bulk parameters $\lambda_K = 1.0$ and $Wi = 1.0$ stand for a kayaking-tumbling (out-of-plane) attractor. By comparison, for a higher Weissenberg number $Wi = 4.0$ (representing tumbling bulk behavior) the principal director shows wagging motion in the entire pore. In [158] the flow modes of in-plane symmetry adapted alignment for a similar model were classified. Within this classification scheme the upper panel of Fig. 4.15 is referred to as tumbling-wagging composite mode with inside defects. The occurrence of defects in the transition layer is typical and will be discussed under full hydrodynamical conditions in the next section.

For higher gap separation to elasticity length ratios ($Er = 1000$) out-of plane solutions become possible. In the Fig. 4.16 the spatio-temporal behavior of the order parameter a for composite out-of plane flow modes are displayed. The upper figure show kayaking-tumbling to kayaking-wagging composite flow mode and the lower panel a composite flow mode that consists of kayaking-wagging in the center of the gap and oscillations in the boundary regions. The oscillating motions in the boundary layer do not fit to any bulk solutions and are rather

4.3. ORIENTATIONAL DYNAMICS AND FLOW PROPERTIES OF NEMATIC STATE

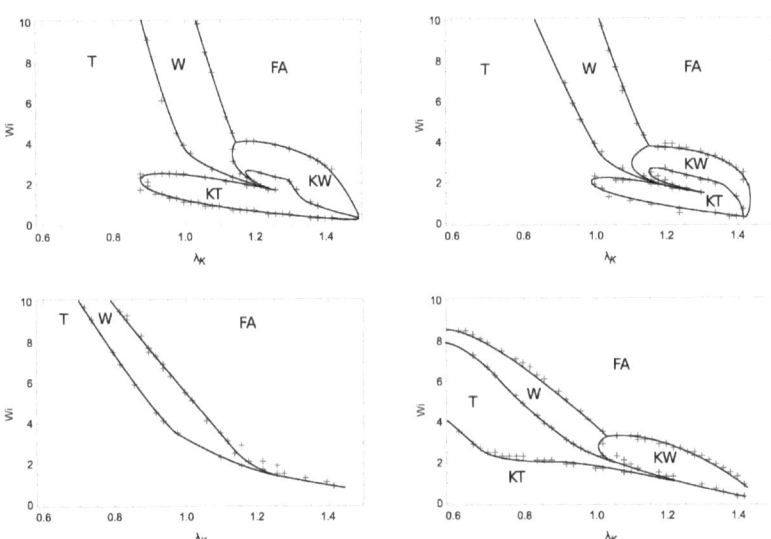

Figure 4.19: Rheological phase portraits for planar anchoring boundary conditions calculated in the center of the Couette cell. For long elastic range the physical conditions at the plates significant effect the orientational behavior inside the cell. The Ericksen numbers are $Er = 1000$ (upper left), $Er = 500$ (upper right) and $Er = 100$ (lower left). In contrast, for isotropic anchoring conditions symmetry breaking states are less suppressed (lower right figure, $Er = 100$). The parameter $\kappa_a = 0$.

irregular. In both flow modes the transition region is characterized by low order parameter values due to increasing biaxiality (hint of defects) and high alignment gradients. For specific parameters $\lambda_K = 1.175$, $Wi = 3.75$ and for a high Ericksen number $Er = 1000$ irregular orientational dynamics in the bulk region set up after a relative long quiet phase (see, Fig. 4.17).

In principle the characteristic flow modes depend on the specific values of the Ericksen number, the tumbling parameter and the Weissenberg number (proportional to the Deborah number). In the case where the Ericksen number and Weissenberg number is increased the system passes through different characteristic solutions. This effect is known as the Deborah number and Ericksen number cascade, respectively and was investigated numerically and found in experiments [61, 163]. The dependence of the Weissenberg number ($Wi = 0.1...10.1$) on the time-averaged order parameter and biaxiality parameter is illustrated in Fig. 4.18. In all cases planar anchoring conditions are used. For low shear rates the biaxiality is close to zero, except of in small boundary regions. When the Weissenberg number is increased, the biaxiality increases up to a maximal value. This effect is more pronounced for high Ericksen numbers and low tumbling parameters.

CHAPTER 4. SPATIALLY INHOMOGENEOUS DYNAMICS OF NON-POLAR HARD-ROD FLUIDS

The dependence of the bulk dynamics on Wi and λ_K was investigated within rheological phase diagrams [44, 50–52, 77]. To investigate the Wi-λ_K scaling behavior for the inhomogeneous system the orientational dynamics in the center of the Couette cell is considered and a rheological phase plot is calculated as follows.

The separation between the plates is divided into ten equidistant regions. For a specific parameter set the five components of the alignment tensor and the tumbling angle for the principal director are calculated. Over each region the data are averaged and the characteristic bulk solution is classified, i.e. isotropic (I), flow alignment (FA), wagging (W), tumbling (T), kayaking-wagging (KW), kayaking-tumbling (KT) and log rolling states (L). For a detailed description of the classification see [51] and references therein. Here a grid of 100×100 points was used to span the two parameter range $\lambda_\mathrm{K} \times Wi = [0.6, 1.5] \times [0.1, 10.1]$.

In Fig. 4.19 the rheological phase diagram for planar anchoring conditions and different Ericksen numbers are presented. Depending on the values of Er the diagram changes strongly. For $Er = 1000$, representing wide plate separation compared to the elastic length ξ_a large parameter ranges of symmetry breaking solutions can be identified. The parameter range for the out-of-plane flow modes shrinks with $Er = 500$ and disappears for $Er = 100$. For long elastic range (low values of the Ericksen number) the physical conditions at the plates significant impose the dynamical behavior in the Couette cell. Here the out-of-plane solutions are suppressed due to the planar anchoring conditions. The lower right figure shows the rheological phase diagram calculated with isotropic anchoring conditions. Despite the low Ericksen number out-of-plane solutions are not suppressed. This indicates that not only the existence of boundaries is important for the orientational behavior but also the specific orientation on the walls.

4.3. ORIENTATIONAL DYNAMICS AND FLOW PROPERTIES OF NEMATIC STATE

4.3.2 Hydrodynamics: Oscillating Jet-Layers

In this section the full nemato-hydrodynamics including flow-feedback is studied. In real systems the flow behavior of complex fluids often shows non-monotonic flow curves (shear vs. stress dependence) as the present model does. Fluids with a non-monotonic flow curve can exhibit two different shear rate values corresponding to one stress value. In the consequence a hydrodynamical instability arises characterized by non linear flow profiles referred to as shear banding. Shear banding phenomena are studied experimentally for e.g. in wormlike micelle solutions and are investigated theoretically by the Johnson Segalman model [16–18, 161, 164–170]. The physical reason for shear banding is the different spatio-temporal microstructure in the fluid (eg. isotropic vs. nematic phase). It is expected that regions with different orientational behavior also show interesting spatio-temporal flow response due to the anisotropy of the viscosity.

The non-Newtonian flow feedback, especially the alignment gradient contributions to the pressure tensor are challenging for the numerical algorithm. In addition to the numerical program used in the previous section new non-standard schemes are implemented (in more detail, see appendix). For the validation of the numerical algorithm solutions are compared to the analytical results from chapter 4 within the graphical precession. Moreover a mesh convergence analysis was performed (see appendix).

In the simulations no significant additional effect (as found in chapter 4) onto the flow profile due to the alignment flux boundary condition was found. High values of the parameter D_a (scales the strenth of the alignment flux) leads to a smooth out the alignment gradients and flow responses. To simplify the analysis the parameter D_a is set to be zero and the parameter c_a that scales the apparent slip to a high value $c_a = 1000$ such that no significant apparent slip associated to the alignment flux boundary condition is considered. To focus on the non-Newtonian flow response on composite solutions, the parameter values for tumbling-wagging composite solutions $\lambda_K = 1.0$, $\vartheta = 0$, $\kappa_a = 0$, $Wi = 1.0$ and $Er = 100$ are used. In Fig. 4.20 the time-dependence of the flow profile is given for the hydrodynamical parameters $\beta = 1.0$ and $\nu_{iso} = 0.1$. The flow response shows non-monotonic flow profiles characterized by oscillating local spurts and a time-dependent apparent velocity slip. Here, the apparent velocity slip is not a consequence of the alignment flux absorption into the wall. The underlying reason is different. Close to the boundary the orientation is parallel to the plates (planar boundary conditions). Due to the Miesowicz viscosities the boundary layer is more viscous and this leads to an apparent slip. For homeotropic anchoring conditions the boundary layer is less viscous that implies to a stick-like flow [170]. The spurts are oscillating with the tumbling period and occur in the tumbling-wagging transition region.

4.3.3 Oblate Defects and Jet-Generation Mechanism

The non-Newtonian flow feedback effect is closely related to biaxial orientational configurations. To capture the tensor character of the orientational dynamics, scientific visualizations via ellipsoids are used (explained in the first chapter).

In Fig. 4.22 the time dependence of the alignment tensor represented by ellipsoids is shown. In the center of the gap the orientational behavior is comparable with bulk tumbling motion.

CHAPTER 4. SPATIALLY INHOMOGENEOUS DYNAMICS OF NON-POLAR HARD-ROD FLUIDS

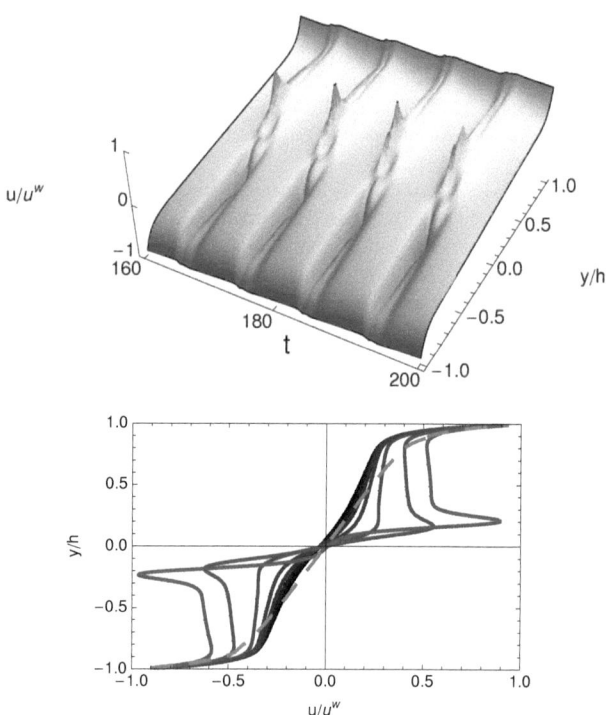

Figure 4.20: Time-dependent velocity profile for the parameters $\vartheta = 0$, $\beta = 1.0$, $\nu_{\text{iso}} = 0.1$, $\kappa_{\text{a}} = 0$, $Wi = 1.0$, $\lambda_{\text{K}} = 1.0$, $Er = 100$, $a_{\max} = 2.5$. Planar anchoring conditions were used. In the lower panel snap shots of the flow showing the generation a jet-like layer is shown for times $t = 182.0 - 183.4$.

4.3. ORIENTATIONAL DYNAMICS AND FLOW PROPERTIES OF NEMATIC STATE

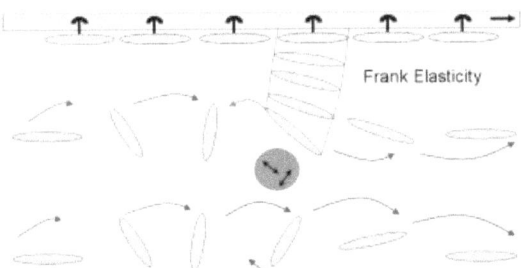

Figure 4.21: In the sketch the formation of oscillating plate-like defects is illustrated. Plate-like defects are a result of the competition between the Frank elasticity, strong anchoring conditions (wagging in the boundary layer) and a torque on the director due to the velocity gradient (tumbling in the middle region). The mismatch between tumbling and wagging leads to plate-like defects.

The regions close to the plates are characterized by wagging motion (as for the wagging-tumbling composite solution for the imposed shear flow). In wide regions and for long times the ellipsoids are cigar shaped. For cigar shaped ellipsoids the orientation is close to uniaxial alignment and a principal director can be specified. However, in the transition layer the ellipsoids become plate-like. Here two directions are equal corresponding to a planar biaxial orientation. This special case is referred to as oblate-defect (plate-like defect). The oblate defect has to be distinguished to topological defects (in one spatial dimension topological defects cannot occur). In the lower panel of the figure 4.22 at time $t = 86.2$, and position $y = 0.2$ an oblate defect is shown. Due to the symmetry of the system two transition regions existing (two walls), oblate defects are creating and annihilating pairwise.

The figure 4.21 shows a sketch of the oblate defect formation and annihilation mechanism. The strong anchoring condition fix uniaxial alignment at the plates. Due to the Frank elasticity it costs energy to deviate from the uniaxial orientational configuration. The length scale of the boundary-elasticity effect is given by the Ericksen number (high range means low Ericksen number). On the other hand, the viscous torque of the sheared fluid forces the principal director into a tumbling motion. The mismatch of the two topologically different bulk dynamic states (tumbling, wagging) results in the creation of oblate defects. The oblate defect is able to adept the wagging principal director to the tumbling principal director.

In Fig. 4.23 the principal director angle in the tumbling and wagging layer is displayed, respectively. In the middle of the gap the tumbling motion is smooth. Close to the oblate defect the principal directors in the tumbling and wagging regime move in phase up to a critical point. At the critical point the directors are stopping for a time. The wagging director flips back and the tumbling forward such that both are again in phase. The jet emerges in the defect-layer at the same time when the directors are stopping and the oblate defect is created. At this time the local orientational configuration favors low local viscosities that leads to the local spurt of the velocity.

In classical experiments of Kiss, Porter and Cementwala [171, 172] unusual sign changes

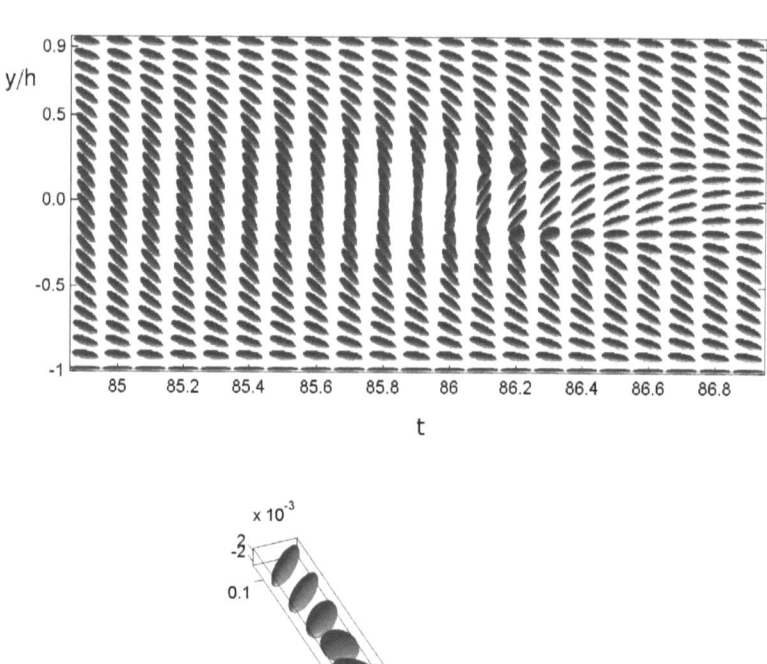

Figure 4.22: The upper figure show the time evolution of the alignment tensor visualized by ellipsoids. At the time $t = 86.2$ a plate-like defect pair is created. The lower figure is a zoomed version of the upper figure. The model parameters are $\vartheta = 0$, $Wi = 1.0$, $Er = 100$, $\lambda_K = 1.0$, $\kappa_a = 0$, $\beta = 1.0$ and $\nu_{iso} = 0.1$.

4.3. ORIENTATIONAL DYNAMICS AND FLOW PROPERTIES OF NEMATIC STATE

of the normal stress differences were observed in experiments. It could be shown that sign change is typically associated with the tumbling transition. Here the normal stress differences show layers of pulsating positive and negative values that coincides with the jet-oblate defect layers (see, Fig. 4.23).

4.3.4 Multiple Jets and Scaling Behavior

To investigate the dependence of the flow dynamics on elastic gradients the simulations are extended to higher values of the Ericken number (all other parameters are not changed). As a result a more complex pulsating jet-band structure with three jet pairs across the gap has been found (see, Fig. 4.25). In the lower panel of Fig. 4.25 the spatio-temporal behavior of the order parameter a shows sharp peaked low values that indicate oblate defects. The orientational configuration in the defects is not always exactly planar biaxial as discussed in the previous section and the local spurt effects are less pronounced. For higher Ericksen numbers the number of defects increase whereas the sharpness (the orientational gradients) decrease. In fact, higher Ericksen numbers yield to smaller elastic length (constant plate separation). The elastic length sets a length scale to the cost of elastic energy due to the distorsion of the alignment tensor field. As a result, for small elastic length small "domains" with rather uniform orientation are set up (large for large elasticity length) leading to more "domain walls". The thickness of the domain walls seems to be shrinking with increasing Ericksen number.

The underlaying mechanism of the jet creation close to the plates is the same as elucidated in the previous section. However, defects located in the middle region of the Couette cell are created differently. In that region the principal director motion is always tumbling, but with a different periodicity. This implies a mismatch and oblate defects are generated.

As the lower panel of Fig. 4.25 indicates, the apparent slip is smaller when compared with the velocity profile in figure 4.20. Due to the higher Eriksen number, the boundary layer (low viscosity) is much smaller and therefore the apparent slip is less pronounced.

As already mentioned, the location of the tumbling-wagging transition interface is determined by the elasticity length ξ_a. In Fig. 4.26 the Wi and Er scaling behavior of the boundary (wagging) layer thickness is illustrated, respectively. The thickness ζ is proportional to the square root of the Eriksen number that means proportional to the elasticity length. The nonlinear fit gives

$$\zeta = 3.9 \left(\gamma^{\text{eff}}\right)^{-1} \frac{\xi_a}{2h}. \tag{4.101}$$

The dependence of ζ on the Weissenberg number is different. High shear rates yield larger tumbling regions, that is in agreement to the wagging bulk behavior. The high error bars between $Wi = 2...3$ are possibly due to the bistability known from bulk solutions. The non-Newtonian flow feedback phenomena reported here could be confirmed by the Doi-Marrucci-Greco and the kinetic models (see [88, 173] for the collaborating work).

CHAPTER 4. SPATIALLY INHOMOGENEOUS DYNAMICS OF NON-POLAR HARD-ROD FLUIDS

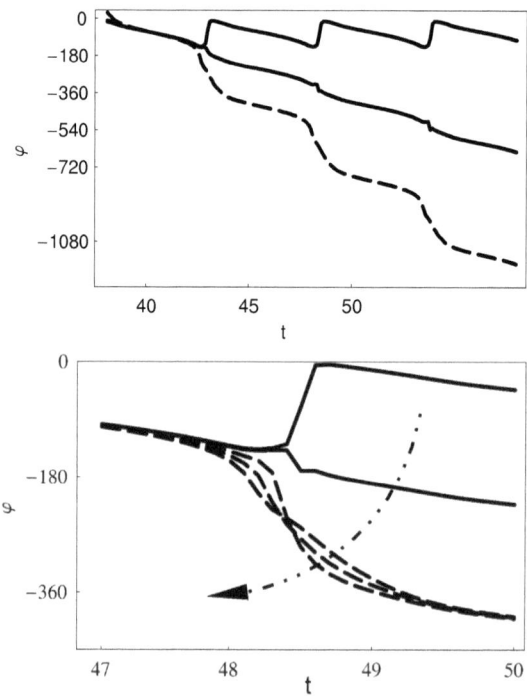

Figure 4.23: The time dependence of the Leslie angle (angle between the principal director projected onto the shear plane and the flow direction) for different layer positions. The upper panel gives the angle at $y = 0.21$ (thick line), $y = 0.20$ (thin line) and $y = 0.0$ (dashed line). The lower panel is a zoomed in version of the upper one for $y = 0.21$, $y = 0.2$, $y = 0.15$, $y = 0.19$, $y = 0.0$ (along the arrow). The parameters used as in in Fig. 4.22.

4.3. ORIENTATIONAL DYNAMICS AND FLOW PROPERTIES OF NEMATIC STATE

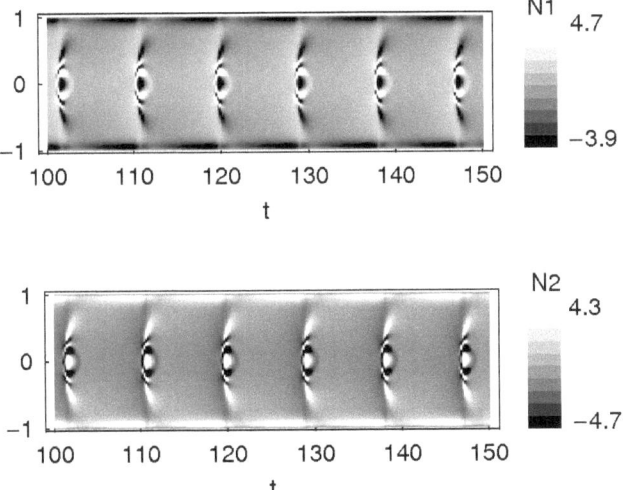

Figure 4.24: The first and second normal stress differences show pulsating negative and positive values that coincide with oblate defects. The parameters are the same as used in the Fig. 4.22.

CHAPTER 4. SPATIALLY INHOMOGENEOUS DYNAMICS OF NON-POLAR HARD-ROD FLUIDS

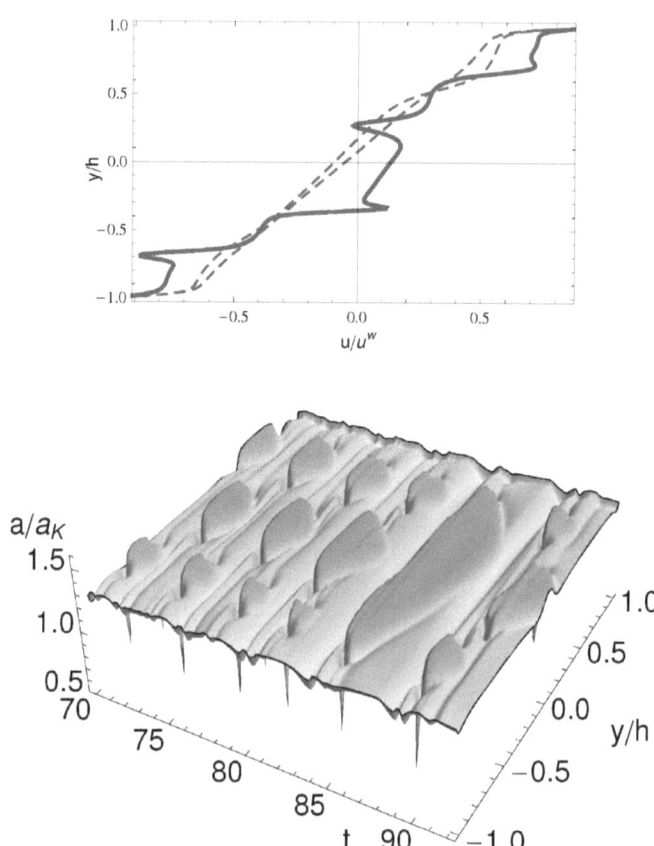

Figure 4.25: Upper panel: The velocity profile for the time $t = 88.5$ shows three jet pairs. Lower panel: The spatio-temporal dynamics of the order parameter a show up to three pairs of defects at the same time. The kayaking-tumbling bulk behavior parameter are used, i.e. $\lambda_K = 1.0$, $Wi = 1.0$, $\kappa_a = 0.0$. The hydrodynamical parameters are $\beta = 1.0$, $\nu_{\text{iso}} = 0.1$, and the Ericken number is $Er = 800$.

4.3. ORIENTATIONAL DYNAMICS AND FLOW PROPERTIES OF NEMATIC STATE

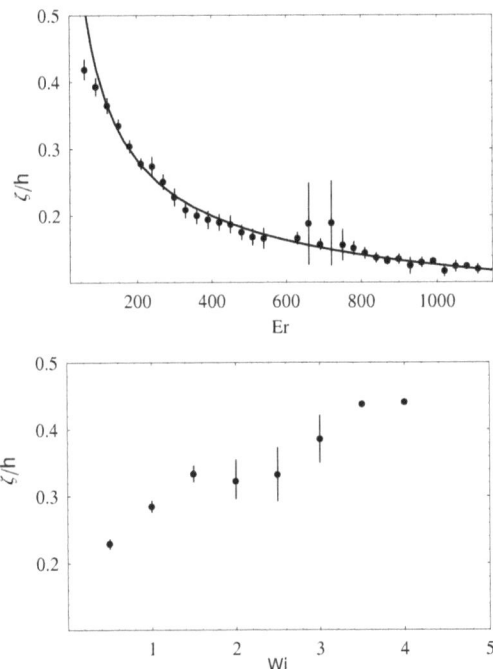

Figure 4.26: Upper panel: The distance between the wall and the average value of the location of the nearest hydrodynamical jet ζ is plotted vs the Erickson number Er. The points are the averages (500 time units) and the lines are the error bars. The curve is the best fit of the function $\alpha/\sqrt{Er}$. The parameters are $\vartheta = 0$, $\kappa_a = 0$, $\beta = 1.0$, $\nu_{\text{iso}} = 0.1$, $Wi = 1.0$, $a_{\max} = 2.5$. Lower panel: The distance ζ is plotted vs the Deborah number De. The parameter are as in the upper panel, except of $Er = 200$.

5
Spatially Inhomogeneous Dynamics of Polar Hard-Rod Fluids

5.1 Shear-Induced Dynamic Polarization and Mesoscopic Structure

The orientational dynamics of rod-like particles with permanent (electric or magnetic) dipole moment in the *bulk* subjected to a shear flow was investigated by Grandner [50–52] The equilibrium nematic to ferronematic phase diagram depending on the dipole coupling constant c was calculated and discussed. The non-equilibrium bulk flow states are classified and a rheological phase diagram (De vs. λ_K) was determined. New transient and in-plane chaotic states could be identified. It turned out that unpolarized equilibrium states show no polarization under shear. In the following chapter spatially inhomogeneous polar rod fluids are investigated. It turns out that spatially inhomogeneous polar rod fluids subjected to a shear flow show steady and dynamical polarizations even when the equilibrium state is non-polarized [174].

In the following the bulk parameters were set to model non-polarized nematic phase at equilibrium, i.e. $c = -30$, $E = 100$, $\vartheta_d = 100$, $\vartheta = 0$, see [51]. The negative value of the dipole coupling parameter indicates a dipole moment preferably parallel to the molecular backbone axis [50, 52]. As in the previous chapter no slip boundary condition for the velocity and different conditions for the alignment (planar, homeotropic, isotropic and degenerate, see 4.100) are used. The boundaries are described in more detail in the previous section. Similar to the calculations in the previous section the initial values are used with consistent to the boundaries. After a relaxation time of 10 time units the flow is switched on by moving the plates.

In Fig. 5.1 the stationary solution for the order parameter $a = \sqrt{\mathbf{a} : \mathbf{a}}$ and the the magnitude of the polarization $|\mathbf{d}|$ for narrow plate separations compared to the elasticity length ξ_a ($Er = 100$) is displayed. No polarization in equilibrium (switched off shear flow) was found. But, after the flow is switched on a time-dependent polarization is set up. The polarization peaks are correlated with high gradients of the order parameter a. A director analysis of the order parameter reveals that the high gradients are due to local pulsating regions of planar biaxial order (oblate defects) as discussed in the previous chapter.

The oscillating defects gives rise to high gradients in the alignment that couples back to

CHAPTER 5. SPATIALLY INHOMOGENEOUS DYNAMICS OF POLAR HARD-ROD FLUIDS

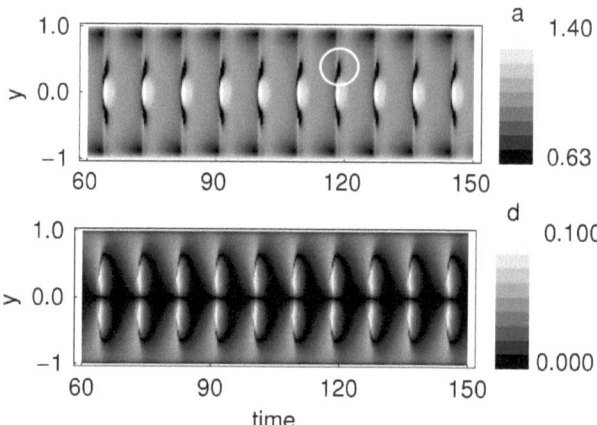

Figure 5.1: The order parameter a (upper graph) and the magnitude of the polarization d (lower graph) is displayed for the model parameters $\lambda_{\rm K} = 1.0$, $Wi = 1.0$, $\mathcal{E}_{\rm fa} = 10$, $\mathcal{E}_{\rm fd} = 10$, $Er = Er_{\rm d} = 100$, $\nu_{\rm iso} = 0.1$, $Wi_{\rm d} = 2$, $\beta = 1$.

the velocity, and to due to the flexoelectricity, to the polarization. For nonpolar hard rods it was shown in the previous chapter that a jetting like behavior of the velocity occurs (the same "nonpolar" parameters are used). The plane Couette flow profile strongly distorted by pairs of pulsating local spurts in the opposite direction. In the presence of polarization the pulsating jet-like layers amplify the polarization.

The relation between orientational structure and flow profile at some selected times is illustrated in Fig. 5.2. In $u(y)$ one observes the formation of pronounced spurt-like deviations from the ordinary linear flow profile (jets), at specific times corresponding to defect oscillations in $a(y,t)$. The occurrence of these jets is a consequence of the feedback effect incorporated in the model [88]. It is seen from Fig. 5.2 that the jets serve as a „trigger" of the local dipole moment $\mathbf{d}(y)$, which becomes particularly large within the spurts.

In Fig 5.3 the magnitude of the polarization vector $|d|$ is shown for different strong anchoring conditions and for lower elastic length ($Er = 300$). Different anchoring conditions strongly affect the spatio-temporal behavior. In the homeotropic case the creation and annihilation of defects pairs is very close to the middle of the gap. This show the wide influence of the boundary condition into the fluid. On the other hand for rough surfaces (isotropic anchoring) high peaks of the polarization occur.

As discussed in the previous section, the boundary layer scales with the Ericksen number and sets a characteristic length for the spatio-temporal structure. High Ericksen number gives rise to complex oblate defect structure and polarization dynamics. In Fig. 5.4 the time

5.1. SHEAR-INDUCED DYNAMIC POLARIZATION AND MESOSCOPIC STRUCTURE

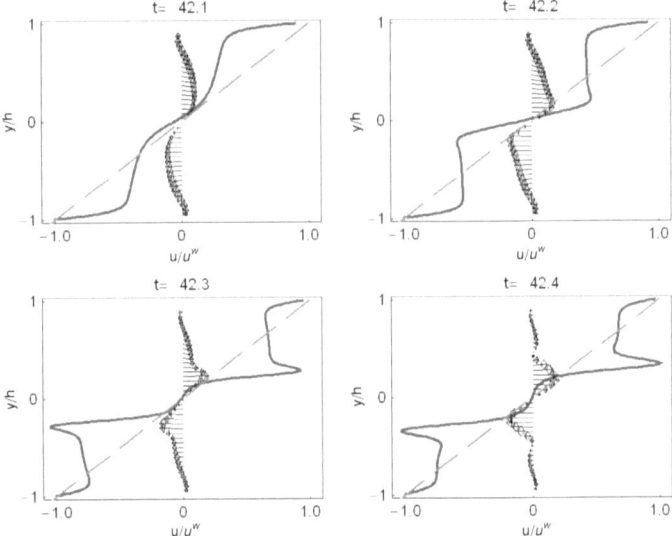

Figure 5.2: Snapshots of the velocity profile and the average dipole vector (arrows). The selected time units ($t = 42.1 - 42.4$) correspond to a time window related to the creation of a defect in Fig. 5.1.

CHAPTER 5. SPATIALLY INHOMOGENEOUS DYNAMICS OF POLAR HARD-ROD FLUIDS

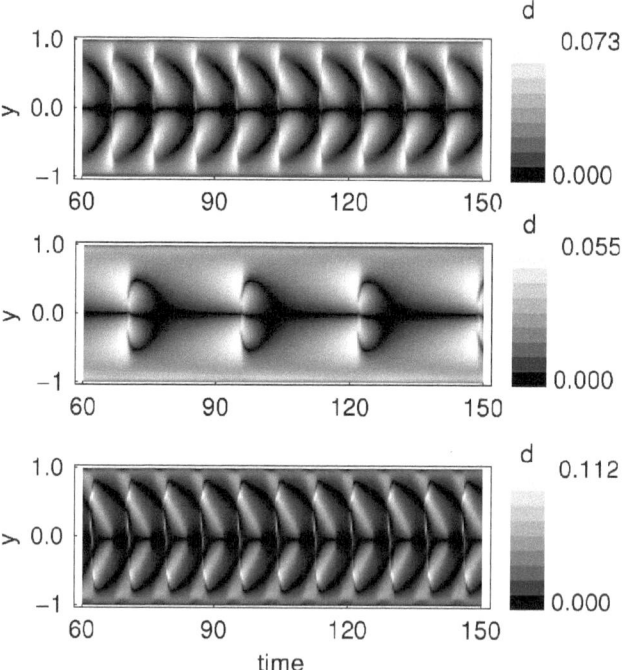

Figure 5.3: The magnitude of the dynamical polarization for monostable planar anchoring (preferred x-direction) in the upper, homeotropic anchoring in the middle and isotropic conditions in the lower is displayed. The figure for the degenerate planar anchoring is qualitatively the same as the upper figure and not shown here. The model parameter used are the same as used in Fig. 5.1, except of $Er = 300$.

5.1. SHEAR-INDUCED DYNAMIC POLARIZATION AND MESOSCOPIC STRUCTURE

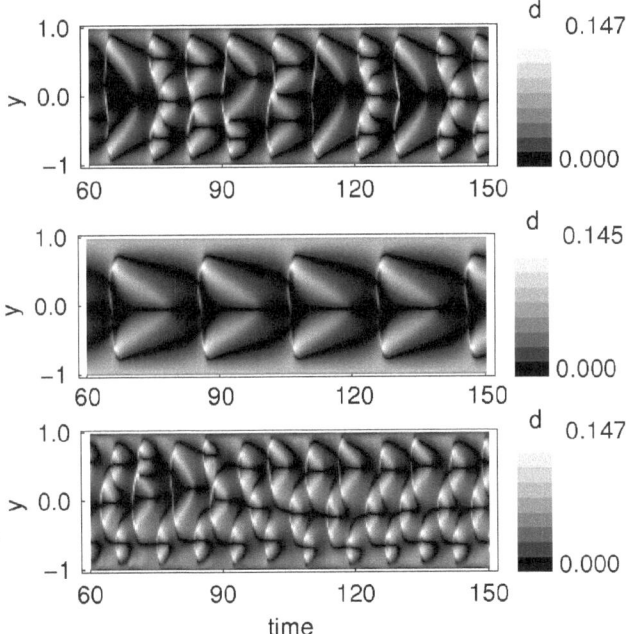

Figure 5.4: The magnitude of the dynamical polarization for monostable planar anchoring (preferred x-direction) in the upper, homeotropic anchoring in the middle and isotropic conditions in the lower is displayed. The figure for the degenerate planar anchoring is qualitatively the same as the upper figure and not shown here. The model parameter used are the same as used in Fig. 5.1, except of $Er = 800$.

CHAPTER 5. SPATIALLY INHOMOGENEOUS DYNAMICS OF POLAR HARD-ROD FLUIDS

evolution of the magnitude of the dipole vector is shown for same parameters and boundary conditions as in Fig. 5.3, except of $Er = 800$. Especially for planar and isotropic boundary conditions multiple oblate defects lead to high spontaneous polarization.

The shear flow induced dynamical polarization gives rise to magnetic fields. The relation between the **B**-field and the polarization was derived in Eq. (2.58). Note, that analogous considerations can be made for systems with magnetic molecular dipoles leading to an electric field. In Fig. 5.5 the time averaged magnetic field in z-direction vs. the flexoelectric coefficient and the Ericksen number Er is displayed, respectively. High values of the magnetic field are expected for high flexoelectric coefficients c_f and high values of Er. The flow-induced polarization disappears for vanishing flexoelectric coupling.

The dependence of the magnetic field on the plate separation and the viscosity are given in Fig 5.5. The higher the viscosity the lower the expected magnetic fields. High viscosities prevent high variations of the velocity profile from the ordinary linear Couette profile and weak polarization is induced. For small plate separation, compared to the elastic length, no textures are generated and only tiny magnetic fields are induced. It follows a range where the elastic length and h is in a ratio where the dynamics generate defect pairs and lattices. The structure contains high elastic gradients and spatially wide ranges are polarized. For high values of h the influence of the wall gets weaker and the gradients are less pronounced. The change of the polarization slows down and the magnetic fields are decreased.

The expected magnetic fields are small. However, in bent-core nematic liquid crystals (C1Pbis10BB) giant flexoelectricity of $62nC/m$ was found [175]. If the plate speed $v^W = 2*10^{-2} m/s$, the relaxation times $\tau_a \approx \tau_d = 0.001$, the number density $\rho = 10^{27} m^{-3}$, the magnitude of polarization $p^{el}|\mathbf{d}| = 1.61^{-28}$, the second Newtonian viscosity $\eta_{iso} = 10 mPs$, the plate separation $2*10^{-2}m$, the elastic lengths $\xi_a = \xi_d = 10^{-5}$, $a_K \approx 1$ and the parameter $\delta_K \approx 0.1$ are used, the scaled time averaged magnetic field is $B_z^* = 5*10^{-3}$ (here $c = 30 > 0$ to model bend-core nematics). The corresponding rescaled field is of the order $O(10^{-11})$, such that the expected magnetic fields are very small. However, the detection limit of a super conducting quantum interference device magnetometer (SQUID) is about 10^{-14} T. The *predicted effect is measurable*. It was shown that the dynamical polarization occur for a wide range of parameters. However, the bulk parameters Wi and λ_K, (determine the characteristic bulk solutions) are not varied so far. In table 5.1 for representative bulk parameter the maximum value of the magnitude of the dipole vector and the averaged magnetic fields are given. It turns out, that in the bulk flow alignment parameters the polarization reaches a steady state. Surprising is the fact that for kayaking-wagging bulk parameter the dynamics is suppressed.

In Fig. 5.6 the magnitude of the dipole vector for representative flow modes is shown. In the upper panel the polarization reaches a steady band like state for flow alignment bulk parameter values. The small polarized bands close to the boundaries are due to the alignment gradients in the boundary layer. In the theoretical description of polar rod fluids two characteristic length scales (ξ_a, ξ_d) occur. In the previous figures the parameters are used such that these scales are of the same magnitude. In the middle panel of Fig. 5.4 different scales are used. The spatio-temporal behavior of the magnitude of polarization show a high and a-chiral behavior due to the fact that the dipoles tray to align together (small value of E_d, long distance effect) The lower panel of the figure show the spatio-temporal dynamics of the magnitude of polarization for irregular (chaotic) orientational dynamics. Therefore, polar

5.1. SHEAR-INDUCED DYNAMIC POLARIZATION AND MESOSCOPIC STRUCTURE

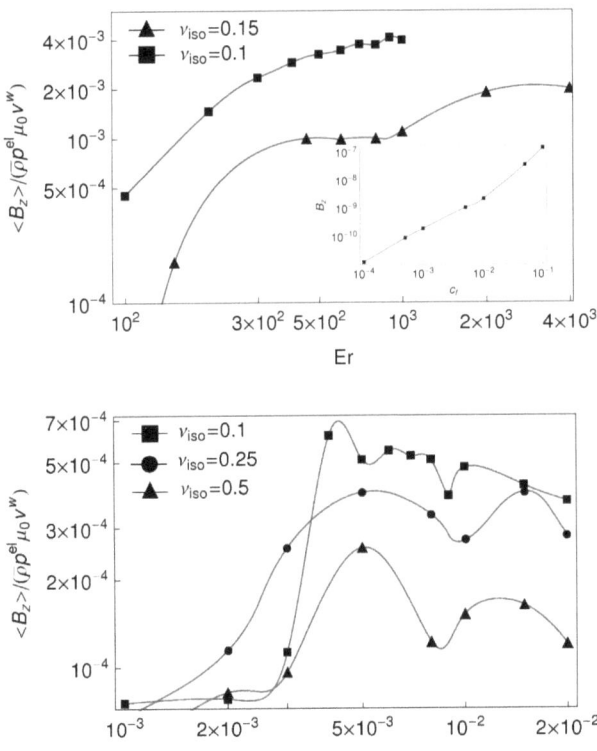

Figure 5.5: The z-component of the time averaged magnetic field B_z vs. the flexoelectric coefficient and the parameter Er is displayed. The other model parameter are chosen as in Fig. 5.1. Here $v^w = 10^{-2} m/s$ and $\tau_a = 10^{-3}s$ is used to give the scaled flexoelectric coefficient.

CHAPTER 5. SPATIALLY INHOMOGENEOUS DYNAMICS OF POLAR HARD-ROD FLUIDS

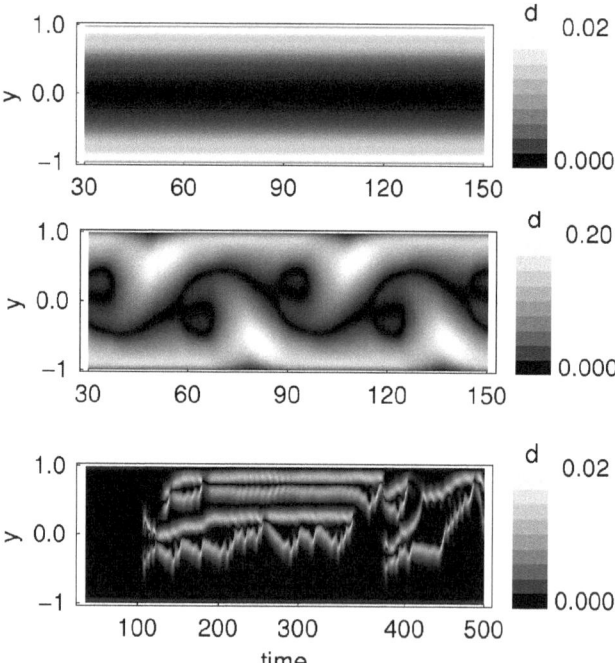

Figure 5.6: The magnitude of the dynamical polarization for monostable planar anchoring (preferred x-direction) is displayed. Upper panel: The magnitude of the polarization for flow alignment bulk parameters shows steady polarized bands close to the plates. The model parameters are $\lambda_K = 1.3$, $Wi = 5.0$. The other parameters are used as in Fig. 5.1 (planar anchoring). Middle panel: The magnitude of the polarization show a-chiral spatio-temporal behavior with high polarization maximum values. The parameters are used as in Fig. 5.1, except of $\mathcal{E}_{fa} = \mathcal{E}_{fd} = 3.3333$, $Er_d = 2$ (homeotropic anchoring). Lower panel: The magnitude of the polarization show irregular (chaotic) behavior. The model parameter values are $Er = 1000$, $\beta = 6.85$, $Wi = 3.7$, $\nu_{iso} = 7.4$, $\lambda_K = 1.12$, $\vartheta_d = 100$, $E = 30$, $c = -2$, $Wi_d = 3.7$, $\mathcal{E}_{fa} = \mathcal{E}_{fd} = 10$.

5.1. SHEAR-INDUCED DYNAMIC POLARIZATION AND MESOSCOPIC STRUCTURE

bulk solution	inhomogeneous system
T ($\lambda_\mathrm{K} = 0.9$, $Wi = 1.0$)	$\bar{B} \neq 0$, $d_\mathrm{max} = 0.023$
W ($\lambda_\mathrm{K} = 1.0$, $Wi = 5.0$)	$\bar{B} \neq 0$, $d_\mathrm{max} = 0.021$
FA ($\lambda_\mathrm{K} = 1.3$, $Wi = 5.0$)	$\bar{B} = 0$, $d_\mathrm{max} = 0.023$
KW ($\lambda_\mathrm{K} = 1.2$, $Wi = 4.0$)	$\bar{B} = 0$, $d_\mathrm{max} = 0.023$
KT ($\lambda_\mathrm{K} = 1.0$, $Wi = 1.0$)	$\bar{B} \neq 0$, $d_\mathrm{max} = 0.073$

Table 5.1: Maximal polarization d_max and presence of magnetic fields $\bar{B}$ at $\tilde{E} = 300$ and parameters (W, λ_K) corresponding to bulk tumbling (T), wagging (W), flow alignment (FA), kayaking-wagging (KW), and kayaking-tumbling (KT) as introduced in section (3.1). The remaining parameters are as in Fig. 5.1.

hard-rod fluids reveal rheo-chaos even when the polarization in domains the main director select one direction.

6

Summary, Conclusions and Outlook

6.1 Summary and Conclusions

The orientational behavior and flow properties of polar and non-polar hard-rod fluids subjected to a flow were investigated. The effect of the presence of walls onto the orientational and flow dynamics was studied. The full alignment tensor (orientation in three dimension) was considered.

To simplify the complexity of the nemato-hydrodynamical equations, the present work was focused on the treatment of spatially one-dimensional system.

In the second chapter the theoretical description of polar and non-polar hard-rod fluids are introduced. Based on the framework of irreversible thermodynamics the spatially inhomogeneous relaxation equations for the dipole vector and the alignment tensor are derived. The constitutive equation for the pressure tensor with dipolar and orientational contributions are presented.

Before the spatially inhomogeneous system was studied, the homogeneous system was analyzed in the third chapter. In the theoretical description of the isotropic-to nematic phase transition of hard-rod fluids the Landau-de Gennes potential was successfully used over decades. However, the Landau-de Gennes potential does not restrict the order parameter. In fact, this becomes problematic for the numerical investigation of spatially inhomogeneous systems and especially for modeling polar hard-rod fluids. To cure the problem a new potential (amended potential) to model the nematic-to isotropic phase transition was presented. The potential has the advantage to restrict the order parameter to physically admissible values, coincides with the Landau-de Gennes potential for small values of the order parameter and can be derived from microscopic principles using the Onsager's excluded volume potential. The amended model potential leads for shear flows to the same characteristic solutions at slightly shifted model parameters λ_K and Wi. This can be explained by the difference of the equilibrium values of the order parameter a_{eq}. The equilibrium value is given by the condition $\Phi'(a_{eq}) = 0$ and therefore depends on the value of a_{max}.

In contrast to plane Couette flow, extensional flows lead to different results. In the case of four-rolls-mill flow it could be shown that the amended potential prevents the order parameter to increase boundless. The scalar order parameter values calculated with the relaxation equation including the amended potential coincide with experimental results very well. For further investigations the amended potential was used.

CHAPTER 6. SUMMARY, CONCLUSIONS AND OUTLOOK

In spatially inhomogeneous systems the shear rate is no longer constant in time and space. In order to study the effect of time-dependent derivations of the shear rate on the bulk orientational dynamics the response of the bulk system onto time-dependent shear rate perturbations were considered. It was found that without exceptions the bulk dynamics is very robust against shear rate perturbations. Only small distorsions on the trajectory could be observed. Especially, it was surprising that even chaotic solutions are very sensitive against the change of averaged shear rate but not against perturbations. The study of the largest Lyapunov exponent shows that no small parameter values ξ (perturbation strength) exist such that the chaotic behavior is lost. According to the robustness of the bulk characteristic solutions it may be expected that in inhomogeneous system similar orientational behavior (especially for wide plate separations) can be found. Indeed, beside new kind of orientational behavior in spatially inhomogeneous systems steady (FA), oscillating (KT, KW, T, W) and its composite flow modes are found [41, 57–59, 158]. Even related to the bulk chaotic solutions rheo-chaos (chaotic spatio-temporal dynamics) was found [16, 24, 54, 56, 161].

In chapter 4, the flow properties of the inhomogeneous system of non-polar hard-rod fluids in the isotropic phase subjected to weak flows was investigated. The boundary conditions for the alignment flux tensor are derived within the framework of non-equilibrium thermodynamics and lead to an apparent slip in different flow geometries. In a plane Couette flow as well as in a cylindrical Couette flow, the velocity in the bulk extrapolated towards the wall is slower than the wall velocity. The cylindrical Couette flow is influenced by the radial geometry leading to non symmetrical flow profile, but qualitatively the effect of the alignment flow at the boundary give the same apparent slip and coincides in the limit for large radii with the plane Couette flow profile. On the other hand, for Poiseuille flow and the flow down an inclined plane, the flow in the bulk becomes faster caused by boundary effects. It has been shown for the case of a plane Couette flow and a plane Poiseuille flow, that as a consequence of the alignment flow boundary conditions, the effective viscosity decreases if the length scale (h) of the device is comparable with the slip length. This also applies to the other geometries.

In all cases the boundary effects are dramatic for values of h comparable with the length ℓ. As known from microfluidics, the flow behavior can be strongly affected by boundary conditions and the findings are presented here are highly relevant for micro-fluidic devices. On the other hand, if the systems length is much larger than ℓ, all these boundary effects are negligible, as expected.

The microscopic interpretation of the boundary conditions is a challenging problem in particular for complex fluids. The strong slip limit $h \ll \ell$ of the expressions given here should be taken with caution. Based on experience with rarefied gas dynamics, strong deviations from hydrodynamics require modified differential equations and additional boundary conditions which are not considered here. For simple fluids, a lower bound on h/ℓ for a hydrodynamic description has been inferred from molecular dynamics simulations in [176].

In the second part of the 4th chapter the orientational behavior and flow properties of non-polar hard-rod fluids under shear are investigated. An interesting and surprising flow feedback effect was found. Pulsating jet layers arise in the model, where the flow profile across the shear gap is non-monotonic. The jet layer forms precisely at the location and time when neighboring orbits of the orientational distribution transition from tumbling to wagging, i.e., monotonic rotation versus oscillation in the principal director. The orientational distribution

6.1. SUMMARY AND CONCLUSIONS

forms an oblate defect phase that coincides with the jet layer "pulse", which occurs when nearby tumbling and wagging orbits develop director phase incoherence. When the local axes of orientation regain phase coherence, the flow returns to monotonic shear profiles.

In [88] the pulsating jet layer-oblate defect phase phenomenon is also be found in the numerical analysis of two further models (Doi-Marrucci-Greco and kinetic model). In addition the authors of [88] show the robustness of the phenomenon against two different constraints. First, the tangential anchoring of the orientational distribution at the plates was shifted to an out-of-plane boundary condition. The result is a composite spatio-temporal attractor which exhibits kayaking in the middle of the shear gap and tilted kayaking in layers buffering each plate. The pulsating jet layers form precisely at the transition from monotone major director rotation (kayaking) to finite oscillations (tilted kayaking). Furthermore, oblate defect phases form, correlated in space and time with the pulsating jet, when the local orbits lose phase coherence. Finally, it could be shown that the jet-effect is indeed one-dimensional even the system is perturbed in two-dimensions.

In the last part of this chapter the simulations were extended to find a more complex pulsating shear band structure, with four jets across the gap. Thus alternating layers of wagging and tumbling were found, with jets and oblate defect phases forming at the layer transitions. These phenomena are reminiscent of defect gas-like textures in [177] (without flow feedback).

The formation of composite tumbling-wagging 1D heterogeneous attractors have been observed previously, including studies of Tsuji and Rey [59] and the authors of [98], where in-plane symmetry was imposed and pure shear was also imposed. The present simulations do not enforce in-plane symmetry, and solve for the fully coupled flow, yet it was found that the space-time attractor is in-plane; the out-of-plane degrees of freedom in the tensor and full kinetic distribution function simply decay to zero. The conditions on Wi and Er simply have to be tuned to amplify the flow feedback jet phenomenon where the flow profile becomes non-monotonic and jet-like layers arise. Due to the small Ericksen number used in the simulations the results are mostly applicable in microfluidic devises where the predicted effects are expected and could be measured in the near future.

For the investigation of polar hard-rod fluids the model equations for two spatially inhomogeneous tensorial order parameters, the dipole vector (modeling average dipole moments) and the alignment tensor are derived within the framework of non-equilibrium thermodynamics (chapter 2). For further investigations of ferronematics and for different theoretical approaches see [178–185]. A wide range of materials show induced polarization by orientational distorsions referred to as flexoelectric effect. The flexoelectric effect couple the alignment tensor gradients to the dipole vector order parameter and gives rise to interesting orientational phenomena [186]. The model equations presented here capture the flexoelectric phenomenon that yields surprising and exciting flow and polarization effects.

The last chapter deals with the hydrodynamical properties of polar flexoelectric and non-flexoelectric hard-rod fluids. For non-flexoelectric polar hard-rod fluids no shear-induced polarization was detectable. However, flexoelectric hard-rods fluids show spontaneous dynamical and steady polarization that generates magnetic fields. The hydrodynamic model for structured suspensions of polar hard-rod fluids predicts spontaneous time-dependent polarization. This novel and surprising flow effect is robust against details of the wall boundary

CHAPTER 6. SUMMARY, CONCLUSIONS AND OUTLOOK

conditions. What is essential, however, is a sufficiently large coupling between local dipole moments and alignment gradients, small plate separation compared to the elastic length scale, and a not too large viscosity allowing for an efficient feedback effect.

For special parameters the magnetic fields are of an measurable magnitude. The bulk parameters are to be prepared according to the tumbling behavior (as shown in [32] for colloidal nano-rods and in [31] for liquid crystals). In the present study it was assumed that the mesoscopic structure of the sheared fluid is induced by walls (disregarding local confinements effects such as layer formation [187]).

The polarization effect discovered here may have potential for several applications. For nano-rods, measuring the resulting magnetic or electric field would be a convenient way to detect and control the „flexoelectric" feature in a fabrication process. Furthermore, our predictions offer a way to generate, in a controlled manner, time-dependent fields by streaming polar nano-rods through microfluidic devices [188]. This is similar to the electrokinetic effect used to investigate surface potentials and to construct microchannel batteries [189].

6.2 Outlook for Further Investigations

In principle the relaxation equations capture the structure of dynamical Landau-Ginzburg equations [190] for two order parameters with the symmetry character of "spin one" (dipole vector) and "spin two" (symmetric traceless alignment tensor of rank two). The description is rather general and applicable for a wide range of materials characterized by this symmetry. An extension for order parameters with a higher spin-symmetry character was investigated by Hess [191, 192]. It would be interesting to apply this approach to the recently synthesized tetrapod-nano crystals [193].

A remarkable hydrodynamical instability forming non-linear flow profile is shear banding. The present models gives a non-monotonic relation between shear stress and shear rate [47] (as does the frequently used Johnson-Segalman model [17, 18]) and thus, the essential ingredient for shear bands. Their occurrence and interplay with the spontaneous polarization observed here as well as with oblate defects will be challenging and interesting work for the future.

Furthermore, by including "activity terms" to the constitutive pressure tensor the present model equations (1.65) can be extended to polar and apolar active biomaterials [194–206].(active nematics). The complex flow behavior of active bio-materials (such as bacterial swimmers and actomyosin solutions) is exciting, surprising and characterized by its internal orientational degree of freedom. For example it can exhibit *spontaneous* flow even in the absence of an external force [207–209]. It is expected, that the effects found in this work has novel counterparts in the flow behavior of active nematic bio-materials.

7 Appendix

7.1 Numerics

Bulk dynamics

The equations modeling the bulk orientational dynamics are a coupled system of five first order nonlinear ordinary differential equations. The solutions were derived by numerical integration. The forth order Runge-Kutta integrator, and for time adaptivity, the Runge-Kutta-Fehlberg algorithm were used. To model shear rate perturbations the system is perturbed at every time $\triangle t$ by Gaussian distributed (uncorrelated) random numbers. Note, in the case where fluctuations are at the time scale of the dynamics one has to apply stochastic integration methods that are not used here.

Inhomogeneous System

The spatially inhomogeneous hydrodynamical system modeling non-polar hard-rod fluids in one spatial dimension consists of six coupled nonlinear partial differential equations. The numerical solutions are found as follows. The spatial variable of the system is discretized by a forth order finite difference scheme [210, 211]. For the boundary terms asymmetric stencils are used. As a result a system of nonlinear ordinary differential equations is obtained (mostly 800 but up to 2000, depending on the grid discretization). For the numerical solution the system is integrated by a time adaptive forth order Runge-Kutta integrator. In addition, to handle the nonlinear terms and to make the algorithm more stable, nonstandard methods are used as introduced in [212].

To validate the numerical program numerical solutions are compared with the analytical solutions of chapter four within graphical precision. Furthermore, different integrators are used (e.g. Adams-Bashforth-Moulton method) and solutions are compared. In the end a numerical mesh analyses was performed. The error for different mesh sizes is given by $\epsilon_u := ||u^{i+1} - u^i||_2$. Here f^i is referred to the function f on the grid i and u^{i+1} on a grid $i+1$, respectively. The symbol $||...||_2$ indicates the L_2-Norm.

In the Fig. 7.1 the error of the velocity u and the order parameter a is given for several grids. With finer mesh sizes the error tends to zero. At the mesh $i = 800$ the error is of the order $O(10^{-4})$. In the most simulations $h = 0,00125$ was used. At some parameter ranges (as high Ericksen numbers) the differential equations become rather stiff and very fine grids have to be used. In the simulations the error for the integrator was between 10^{-8} and 10^{-13}.

CHAPTER 7. APPENDIX

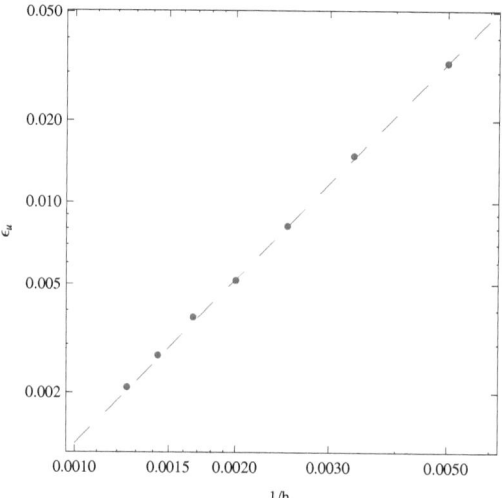

Figure 7.1: Convergence rate for the velocity u vs. the inverse mesh distance h.

7.2. THE PROBABILITY DISTRIBUTION FUNCTION FOR POLAR HARD-ROD FLUIDS

The numerical solutions of the spatially inhomogeneous system modeling the hydrodynamics of polar nano-rods is calculated similarly.

7.2 The Probability Distribution Function for Polar Hard-Rod Fluids

In section 2.1 the distribution function depending on the dipole moment and the molecular axes was introduced. Here the form of 2.4 is motivated. In the following it will be shown, that the first terms in the expansion of the function $\rho^{\text{or}}(\Omega)$ can be written in the form

$$\rho^{\text{or}}(\Omega) = \alpha_1 \mathbf{d} \cdot \mathbf{e}(\Omega) + \alpha_2 \mathbf{a} : \overline{\mathbf{uu}}(\Omega). \tag{7.1}$$

Every function $\rho^{\text{or}}(\Omega)$, where $\Omega = (\theta, \varphi, \alpha)$ notes the tree Euler angles, can be expand in a complete set of rotational matrices $D^l_{mm'}$ [89],

$$\begin{aligned}\rho^{\text{or}}(\Omega) &= \sum_{lmm'} \rho^{\text{or}}_{lmm'} D^l_{mm'}(\Omega). \tag{7.2}\\ &= \sum_{mm'} \rho^{\text{or}}_{1mm'} D^1_{mm'} + \sum_{mm'} \rho^{\text{or}}_{2mm'} D^2_{mm'} + \mathfrak{R}\end{aligned}$$

where

$$\rho^{\text{or}}_{lmm'} = \frac{2l+1}{8\pi^2} \int d\Omega D^l_{mm'}(\Omega)^* \rho^{\text{or}}(\Omega). \tag{7.3}$$

Here it was assumed that the rest term $\mathfrak{R}$ is very small and can be neglected. When the average of the function $A(\Omega)$ is defined as $\langle A(\Omega) \rangle_\Omega = \int d\Omega \rho^{\text{or}}(\Omega) A^*(\Omega)$, then the function $\rho^{\text{or}}(\Omega)$ reads

$$\rho^{\text{or}}(\Omega) = \frac{3}{8\pi^2} \sum_{mm'} \langle D^1_{mm'}(\Omega) \rangle_\Omega D^1_{mm'} + \frac{5}{8\pi^2} \sum_{mm'} \langle D^l_{mm'}(\Omega) \rangle_\Omega D^2_{mm'}(\Omega). \tag{7.4}$$

The rotational matrices are characterized by the rotations of the spherical harmonics Y_{lm}. The transformation equations leads to the following useful expressions

$$Y_{lm'}(\omega') = \sum_m D^l_{mm'}(\Omega) Y_{lm}(\omega) \tag{7.5}$$

$$\langle Y_{lm'}(\omega') \rangle_\Omega = \sum_m \langle D^2_{mm'}(\Omega) \rangle_\Omega Y_{lm}(\omega).$$

The orthogonality relation for the spherical harmonics are given by

$$\langle Y_{lm} Y_{l'm'} \rangle_\omega = \int d\omega Y_{lm}(\omega) Y_{l'm'}(\omega) = \delta_{ll'} \delta_{mm'} \tag{7.6}$$

CHAPTER 7. APPENDIX

With Eq. 7.5 and 7.6 the expression the distribution function is rearranged

$$\begin{aligned}
\rho^{\text{or}}(\Omega) &= 3/(8\pi^2) \sum_{mm'q} \left\langle D^1_{mm'}(\Omega) \right\rangle_\Omega \left\langle Y_{1m}(\omega) Y_{1q}(\omega) \right\rangle_\omega D^1_{qm'}(\Omega) \quad (7.7)\\
&+ 5/(8\pi^2) \sum_{mm'q} \left\langle D^2_{mm'}(\Omega) \right\rangle_\Omega \left\langle Y_{2m}(\omega) Y_{2q}(\omega) \right\rangle_\omega D^2_{qm'}(\Omega) \\
&= 3/(8\pi^2) \sum_{mm'q} \left\langle \left\langle D^1_{mm'}(\Omega) \right\rangle_\Omega Y_{1m}(\omega) Y_{1q}(\omega) D^1_{qm'}(\Omega) \right\rangle_\omega \\
&+ 5/(8\pi^2) \sum_{mm'q} \left\langle \left\langle D^2_{mm'}(\Omega) \right\rangle_\Omega Y_{2m}(\omega) Y_{2q}(\omega) D^2_{qm'}(\Omega) \right\rangle_\omega \\
&= 3/(8\pi^2) \sum_{m'} \left\langle \left\langle Y_{1m'}(\omega') \right\rangle_\Omega Y_{1m'}(\omega') \right\rangle_\omega \\
&+ 5/(8\pi^2) \sum_{m'} \left\langle \left\langle Y_{2m'}(\omega') \right\rangle_\Omega Y_{2m'}(\omega') \right\rangle_\omega .
\end{aligned}$$

The spherical harmonics can be related to the set of symmetric traceless tensors. The general expression is given by [90]

$$\overrightarrow{u^l}_\mu = \frac{4\pi l!}{2l+1)!!} Y_{l\mu}, \quad (7.8)$$

where $\overrightarrow{u^l}_\mu$ = denotes the symmetric traceless part of l-fold tensor product of the unit vector **u**. The orientational distribution function in terms of symmetric traceless tensors reads

$$\begin{aligned}
\rho^{\text{or}}(\Omega) &= \frac{3}{8\pi^2} \left(\frac{3}{4\pi}\right)^2 \left\langle \langle \mathbf{u}(\omega') \rangle_\Omega \cdot \mathbf{u}(\omega') \right\rangle_\omega \quad (7.9)\\
&+ \frac{5}{8\pi^2} \left(\frac{5!!}{8\pi}\right)^2 \left\langle \langle \overline{\mathbf{uu}}(\omega') \rangle_\Omega : \overline{\mathbf{uu}}(\omega') \right\rangle_\omega .
\end{aligned}$$

With the definition

$$\left(\langle \mathbf{u}(\omega') \rangle_\Omega \right)_\mu = \int d\Omega \rho^{\text{or}}(\Omega) D_{\mu\nu}(\Omega) u_\nu(\omega) =: d_\nu(\omega), \quad (7.10)$$

the first term can be written as

$$\begin{aligned}
\left\langle \langle \mathbf{u}(\omega') \rangle_\Omega \cdot \mathbf{u}(\omega') \right\rangle_\omega &= \left\langle d_\mu(\omega) D_{\mu\nu}(\Omega) u_\nu(\omega) \right\rangle_\omega \quad (7.11)\\
&= D_{\mu\nu}(\Omega) \left\langle d_\mu u_\nu(\omega) \right\rangle_\omega \\
&= D_{\mu\nu}(\Omega) d_\mu e_\nu = d_\mu e_\mu(\Omega).
\end{aligned}$$

The same argument is true for the second term and the following probability function is obtained

$$\rho^{\text{or}}(\Omega) = \alpha_1 \mathbf{d} \cdot \mathbf{e}(\Omega) + \alpha_2 \mathbf{a} : \overline{\mathbf{uu}}(\Omega), \quad (7.12)$$

where α_1 and α_2 are constants.

Bibliography

[1] O. Lehmann, Wied. Ann. Physik **24**, 23 (1885).

[2] O. Lehmann, Ann. Physik **38**, 398 (1889).

[3] O. Lehmann, Z. Phys. Chem. **4**, 462 (1889).

[4] O. Lehmann, Wied. Ann. Physik **40**, 401 (1890).

[5] O. Lehmann, Wied. Ann. Physik **41**, 525 (1890).

[6] F. Reinitzer, Monatsh. Chem. **9**, 421 (1888).

[7] P. G. de Gennes and J. Prost, *The Physics of Liquid Crystals*, volume 83 of *International Series of Monographs on Physics*, Clarendon Press, Oxford, 1993.

[8] J. Perez-Juste, I. Pastoriza-Santos, L. M. Liz-Marzan and P. Mulvaney, Coordination Chem. Rev. **249**, 1870 (2005).

[9] D. J. Park, D. C. Kim, J. Y. Lee and H. K. Cho, Nanotechnology **17**, 5238 (2006).

[10] W. Zhou, K. Tang, S. Zeng and Y. Qi, Nanotechnology **19**, 065602 (2008).

[11] H. H. Wensink and G. J. Vroege, Phys. Rev. E **72**, 031708 (2005).

[12] G. Kawamura, Y. Y. Yang, and M. Nogami, J. **90**, 261908 (2007).

[13] A. Gopinath, L. Mahadevan, and R. C. Armstrong, Eur. J. Phys. E **13 (3)**, 309 (2004).

[14] B. J. Lemaire, P. Davison, P. Pannine, and J. P. Jolivet, Phys. Rev. Lett. **93**, 267801 (2004).

[15] B. J. Lemaire, P. Davison, P. Pannine, and J. P. Jolivet, Phys. Rev. Lett. **88**, 125507 (2002).

[16] S. M. Fielding and P. D. Olmsted, Phys. Rev. Lett. **92**, 084502 (2004).

[17] S. M. Fielding and P. D. Olmsted, Phys. Rev. E **68**, 036313 (2003).

[18] S. M. Fielding and P. D. Olmsted, Phys. Rev. Lett. **22**, 224501 (2003).

[19] J. Feng, J. Tao, L.G. Leal, J. Fluid Mech. **449**, 179 (2001).

[20] S. Hess, S. Heidenreich, P. Ilg, and M. Kröger, AIP Conf. Proc. **832**, 311 (2006).

[21] G. Marrucci, Macromolecules **24**, 4176 (1991).

[22] O. Hess, C. Goddard and S. Hess, Physica **366 A**, 31 (2006).

[23] R. Ganapathy and A. K. Sood, Phys. Rev. Lett **96**, 108301 (2006).

BIBLIOGRAPHY

[24] B. Chakrabarti, M. Das, C. Dasgupta, S. Ramaswamy and A. K. Sood, Phys. Rev. Lett. **92**, 55501 (2004).

[25] X. Wang, J. Zhuang, Q. Peng, Y. Li, Nature **437**, 121 (2005).

[26] L.-S. Li and A. P. Alivisatos, Phys. Rev. Lett. **90**, 097402 (2003).

[27] M. J. Bierman, K. M. Van Heuvelen, D. Schmeier, T. C. Brunol, S. Jin, Adv. Mat. **19**, 2677 (2007).

[28] X. Peng, L. Manna, W. Yang, J. Wickham, E. Scher, A. Kadavanich and A. P. Alivisatos, Nature **404**, 59 (2000).

[29] C. Gähwiller, Phys. Rev. Lett. **28**, 1554 (1972).

[30] R. G. Larson, *The Structure and Rheology of Complex Fluids*, Topics in chemical engineering, Oxford University Press, 1999.

[31] R. G. Larson and H. C. Öttinger, Macromolecules **24**, 6270 (1991).

[32] M.P. Lettinga, Z. Dogic, H. Wang, J. Vermant, Langmuir **21**, 8048 (2005).

[33] G. Kiss and R. S. Porter, J. Polymer Sci., Polym. Phys. Ed. **18**, 361 (1980).

[34] Z. Tan and G.C. Berry, J. Rheol. **47(1)**, 73 (2003).

[35] M. Grosso, S. Crescitelli, E. Somma, J. Vermant, P. Moldenaers, and P. L. Maffettone, Phys. Rev. Lett **90**, 98304 (2003).

[36] V. Faraoni, M. Grosso, S. Crescitelli, and P. L. Maffettone, J. Rheol **43**, 829 (1999).

[37] Y.-G. Tao, W. K. den Otter, and W. J. Briels, Phys. Rev. Lett. **95**, 237802 (2005).

[38] Y.-G. Tao, W. K. den Otter, and W. J. Briels, J. Chem Phys. **124**, 204902 (2006).

[39] J. Ding and Y. Yang, Rheologica Acta **33**, 405 (1994).

[40] M. G. Forest and Q. Wang, Rheol. Acta **42**, 20 (2003).

[41] T. Tsuji and A. D. Rey, Phys. Rev. E **62**, 8141 (2000).

[42] M. G. Forest, Q. Wang, and R. Zhou, Rheol. Acta **44**, 80 (2004).

[43] G. Rienäcker, M. Kröger, and S. Hess, Phys. Rev. E **65**, 040702(R) (2002).

[44] G. Rienäcker, *Orientational dynamics of nematic liquid crystals in a shear flow*, Dissertation, TU-Berlin, 2000.

[45] S. Hess, Zeitschrift für Naturforschung **30a**, 728 (1975).

[46] S. Hess, Zeitschrift f. Naturforschung **31a**, 1507 (1976).

BIBLIOGRAPHY

[47] C. P. Borgmeyer and S. Hess, J. Non-Equilib. Thermodyn. **20**, 359 (1995).

[48] M. Grosso and R. Keunings, S. Crescitelli, and P. L. Maffettone, Phys. Rev. Lett **86**, 3184 (2001).

[49] S. Hess and M. Kröger, J. Phys.: Condens. Matter **16**, 3835 (2004).

[50] S. Grandner, S. Heidenreich, P. Ilg, S. H. L. Klapp, and S. Hess, Phys. Rev. E **75**, 040701 (R) (2007).

[51] S. Grandner, S. Heidenreich, S. Hess, and S. H. L. Klapp, Eur. Phys. J. E **24**, 353 (2008).

[52] S. Grandner, *Elektrische Polarisation in strömenden Flüssigkristallen*, Diplomarbeit, TU-Berlin, 2007.

[53] R. Bandyopadhyay, G. Basappa, and A. K. Sood, Phys. Rev. Lett. **84**, 2022 (2000).

[54] M. Das, B. Chakrabarti, C. Dasgupta, S. Ramaswamy and A. K. Sood, Phys. Rev. E **71**, 21707 (2005).

[55] R. Ganapathy and A. K. Sood, Phys. Rev. Lett. **96**, 108301 (2006).

[56] M. G. Forest, R. Zhou, and Q. Wang, Multiscale Model. Simul. **6 (3)**, 858 (2007).

[57] A. D. Rey and T. Tsuji, Macromol. Theory Simul. **7**, 623 (1998).

[58] T. Tsuji and A. Rey, J. Non-Newtonian Fluid Mech. **73**, 127 (1997).

[59] T. Tsuji and A. D. Rey, Phys. Rev. E **57**, 5609 (1998).

[60] R. G. Larson and D. W. Mead, Liq. Cryst. **12**, 751 (1992).

[61] R. G. Larson and D. W. Mead, Liquid Crystals **15**, 151 (1993).

[62] D.H. Klein, C. Garcia-Cervera, H.D. Ceniceros, L.G. Leal, ANZIAM J. **46(E)**, 210 (2005).

[63] C. Goddard, O. Hess, A. Balanov, and S. Hess, Phys. Rev. E **77**, 26311 (2008).

[64] O. Hess and S. Hess, Physica **207 A**, 517 (1994).

[65] A. J. Szeri, J. Rheol. **39 (5)**, 873 (1995).

[66] Q. Wang, J. Rheol. **41 (5)**, 953 (1997).

[67] Q. Wang, J. Chem. Phys. **116 (20)**, 9120 (2002).

[68] A. M. Sonnet, P. L. Maffettone, and E. G. Virga, J. Non-Newtonian Fluid Mech. **119**, 51 (2003).

BIBLIOGRAPHY

[69] H. Stark and T. C. Lubensky, Phys. Rev. E **67**, 061709 (2003).

[70] M. R. Lopez-Gonzalez, W. M. Holmes, P. T. Callaghan and P. J. Photinos, Phys. Rev. Lett. **93**, 268302 (2004).

[71] P. Joseph, and P. Tabeling, Phys. Rev. E **71**, 035303(R) (2005).

[72] D. C. Tretheway, and C. D. Mainhart, Phys. Fluids **14**, L9 (2002).

[73] R. Benzi, L. Biferale, M. Sbragaglia, S. Succi and F. Toschi, Math. and Comp. in Sim **72** (2006).

[74] S. P. Meeker, R. T. Bonnecaze, and M. Cloitre, Phys. Rev. Lett. **92**, 198302 (2004).

[75] V. S. J. Craig, C. Neto, and D. R. M. Williams, Phys. Rev. Lett. **87**, 054504 (2001).

[76] C. Pastorino, K. Binder, T. Kreer, and M. Müller, J. Chem. Phys. **24**, 64902 (2006).

[77] G. Rienäcker and S. Hess, Physica **267 A**, 294 (1999).

[78] G. Rienäcker, M. Kröger, and S. Hess, Physica **315 A**, 537 (2002).

[79] G. Rienäcker and S. Hess, Physica **267 A**, 294 (1998).

[80] L. Waldmann, Z. Naturforsch **22a**, 1269 (1967).

[81] H. Vestner, Z. Naturforsch **28a**, 869 (1973).

[82] S. Hess and H.-M. Koo, J. Non-Equilibrium Thermodyn. **14**, 159 (1988).

[83] R. Fetzer, M. Rauscher, A. Münch, B. A. Wagner, and K. Jacobs, Europhys. Lett. **75**, 638 (2006).

[84] H. C. Öttinger, Phys. Rev. E **73**, 036126 (2006).

[85] D. Bedeaux, A.M. Albano and P. Mazur, Physica **82A**, 438 (1976).

[86] A.M. Albano and D. Bedeaux, Physica **147A**, 407 (1987).

[87] R. Kupferman, M. N. Kawaguchi, and M. M. Denn, J. Non-Newtonian Fluid Mech. **91**, 255 (2000).

[88] M. G. Forest, S. Heidenreich, S. Hess, X. Yang, and R. Zhou, J. Non-Newtonian Fluid Mech. **155**, 130 (2008).

[89] C. G. Gary and K. E. Gubbins, *Theory of Molecular Fluids*, volume 1, Oxford University, London, 1984.

[90] F. R. W. McCourt, J. J. M. Beenakker, W. E. Köhler, and I. Kuščer, *Nonequilibrium Phenomena in Polyatomic Gases*, volume 19 of *International Series of Monographs on Chemistry*, Clarendon Press, Oxford, 1991.

BIBLIOGRAPHY

[91] F. R. W. McCourt, J. J. M. Beenakker, W. E. Köhler and I. Kŭščer, *Nonequilibrium Phenomena in Polyatomic Gases, Volume 2: Cross sections, Scattering and Rarefied Gases.*

[92] S. Hess and W. Köhler, *Formeln zur Tensor-Rechnung.*

[93] W. Maier and A. Saupe, Z. Naturforsch. **13a**, 564 (1958).

[94] W. Maier and A. Saupe, Z. Naturforsch. **14 a**, 882 (1959).

[95] W. Maier and A. Saupe, Z. Naturforsch. **15 a**, 287 (1960).

[96] P. Kaiser, W. Wiese, and S. Hess, J. Non-Equilib. Thermodyn. **17**, 153 (1992).

[97] E. F. Gramsbergen, L. Longa and W. H. de Jeu, Physics Reports **4**, 195 (1986).

[98] M. G. Forest, R. Zhou and Q. Wang, Multiscale Model. Simul. **4**, 1280 (2005).

[99] A. Singh, Phys. Rep. **324**, 107 (2000).

[100] S. Hess, Z. Naturforsch. **31a**, 1034 (1976).

[101] S. Hess and P. Ilg, Rhol. Acta **44**, 465 (2005).

[102] S. Hess and P. Ilg, J. Non-Newtonian Fluid Mech. **134**, 2 (2006).

[103] S. Hess, *Electro-Optics and Dielectrics of Macromolecules*, Plenum Press, New York, 1979.

[104] M. Doi, J. Polym. Science **19**, 229 (1981).

[105] C. Schneggenburger, M. Kröger, and S. Hess, J. Non-Newtonian Fluid Mech. **62**, 235 (1996).

[106] C. D. Southern, P. D. Brimicombe, S. D. Siemianowski, B. Jaradat, N. Roberts, V. Görtz, J. W. Goodby and H. F. Gleeson, Eur. Phys. Lett. **82**, 56001 (2008).

[107] G. R. Luckhurst, Nature **430**, 413 (2004).

[108] K. Neupane, S. W. Kang, S. Sharma, D. Carney, T. Meyer, G. H. Mehl, D. W. Allender, S. Kumar and S. Sprunt, Phys. Rev. Lett. **97**, 207802 (2006).

[109] G. Vertogen and W. H. de Jeu, *Thermotropic liquid crystals, Fundamentals*, volume 45 of *Chemical Physics*, Spinger-Verlag, 1988.

[110] I. Pardowitz and S. Hess, J. Chem. Phys. **76 (3)**, 1485 (1982).

[111] M. Kröger and P. Ilg, J. Chem. Phys. **127**, 034903 (2007).

[112] S. R. De Groot and P. Mazur, *Non-Equilibrium Thermodynamics*, North-Holland Publishing Company, Amsterdam.

BIBLIOGRAPHY

[113] S. Hess and I. Pardowitz, Z. Naturforsch. **36 a**, 554 (1981).

[114] P.Ilg, I. V. Karlin, and H. C. Öttinger, Phys. Rev. E **60**, 5783 (1999).

[115] J. L. Ericksen, Trans. Soc. Rheol **5**, 23 (1961).

[116] F. M. Leslie, Arch. Rat. Mech. Anal. **28**, 265 (1968).

[117] S. Chandrasekhar, *Liquid Crystals*, Cambridge, 1992.

[118] O. Parodi, J. Phys. (Paris) **31**, 581 (1970).

[119] D. Forster, Phys. Rev. Lett. **32 (21)**, 1161 (1974).

[120] G. B. Jeffery, Proc. R. Soc. London Ser. A **102**, 161 (1923).

[121] M. Doi, Ferroelectrics **30**, 247 (1980).

[122] J. Feng, C. V. Chaubal and G. L. Leal, J. Rheol. **42**, 1095 (1998).

[123] H. Pleiner, E. Jarkova, H.-W.Müller, and H. R. Brand, Magnetohydrodynamics **37**, 254 (2001).

[124] H. Pleiner, E. Jarkova, H.-W.Müller, and H. R. Brand, J. Magn. Magn. Mat. **252**, 147 (2002).

[125] G. M. Range and S. H. L. Klapp, Phys. Rev. E **69**, 041201 (2004).

[126] G. M. Range and S. H. L. Klapp, Phys. Rev. E **70**, 031201 (2004).

[127] S. H. L. Klapp and F. Forstmann, Europhys. Lett. **38**, 663 (1997).

[128] B. Groh and S. Dietrich, Phys. Rev. Lett. **72**, 2422 (1994).

[129] J. P. Marcerou and J. Prost, Phys. Lett. **66A**, 218 (1978).

[130] A. Kapanowski, Phys. Rev. E **75**, 031709 (2007).

[131] A. Ferrarini, C. Greco and G. R. Luckhurst, J. Mat. Chem. **17**, 1039 (2007).

[132] S. V. Kalinin and S. Jesse, Appl. Phys. Lett. **88**, 153902 (2006).

[133] D. L. Cheung, S. J. Clark and M. R. Wilson, J. Chem. Phys. **121**, 9131 (2004).

[134] P. L. Maffettone, M. Grosso, M.C. Friedenberg, and G.G. Fuller, Macromolocules **29**, 8473 (1996).

[135] R. G. Larson, Macromolecules **23**, 3983 (1990).

[136] G. Marrucci, Rheol. Acta **29**, 523 (1990).

[137] E. V. Alonso, A. A. Wheeler, and T. J. Sluckin, Proc. R. Soc. Lond. A **459**, 195 (2002).

BIBLIOGRAPHY

[138] M. G. Forest and R. Zhou, J. Rheol. **47 (1)**, 105 (2003).

[139] M. G. Forest and Q. Wang, Rheol. Acta **42**, 20 (2003).

[140] M. G. Forest, S. Sircar, Q. Wang and R. Zhou, Phys. Fluids **18**, 103102 (2006).

[141] P. L. Maffettone and S. Crescitelli, J. Non.-Newtonian Fluid Mech. **59**, 73 (1995).

[142] V. Faraoni, M. Grosso and S. Crescitelli, J. Rheol **43 (3)**, 829 (1999).

[143] J. Mewis, M. Mortier, J. Vermant, and P. Moldenaers, Macromolecules **30**, 1323 (1997).

[144] M. G. Forest, R. Zhou and Q. Wang, Phys. Rev. E **66**, 031712 (2002).

[145] Benettin G., Galani L., Giorgilli A., Strelcyn J. M., Meccanica **15**, 9.

[146] M. G. Forest, R. Zhou and Q. Wang, Phys. Rev. Lett. **93**, 088301 (2004).

[147] C.V. Chaubal and G.L. Leal, J. Non-Newtonian Fluid Mech. **82**, 25 (1999).

[148] L. B. C. Cottin-Bizonne, J.-L. Barrat and E. Charlaix, Nature Materials **2**, 237 (2003).

[149] E. Lauga, M. P. Brenner, and H. A. Stone, *Handbook of Experimental Fluid Dynamics*, Springer, 2005.

[150] C, Neto, D. R. Evans, E. Bonaccurso, H.-J. Butt, and V. S. J. Craig, Rep. Prog. Phys. **68**, 2859 (2005).

[151] T. M. Squires and S. R. Quake, Rev. Mod. Phys. **77**, 977 (2005).

[152] N. V. Priezjev and S. M. Trojan, Phys. Rev. Lett **92**, 018302 (2004).

[153] L. Waldmann and H. Vestner, Physica **99 A**, 1 (1979).

[154] Y. Cohen in N. P. Cheremisinoff, *Rheology and Non-Newtonian Flows*, volume 7 of *Encyclopedia of Fluid Mechanics*, Gulf Publishing Company, Houston, 1988.

[155] W. Rehwald, *Elementare Einführung in die Bessel-, Neumann- und Hankel-Funktionen*, S. Hirzel Verlag, Stuttgart, 1959.

[156] P. Kasperkovitz, J. Math. Phys. **21**, 6 (1980).

[157] B. Jérôe, Rep. Prog. Phys. **54**, 391 (1991).

[158] H. Yu and P. Zhang, J. Non-Newtonian Fluid Mech. **141**, 116 (2007).

[159] R. Zhou, M. G. Forest, and Q. Wang, Multiscale Model. Simul. **3 (4)**, 853 (2005).

[160] M. G. Forest, R. Zhou, Q. Wang, X. Zheng, and R. Lipton, IMA Modeling of Soft Matter **141**, 85 (2005).

[161] S. Manneville, J.-B. Salmon, and A. Colin, Eur. Phys. J. E **13**, 197 (2004).

BIBLIOGRAPHY

[162] D. Grecov, Rheol Acta **44**, 135 (2004).

[163] D. H. Klein, L. G. Leal, C. J. Garcia-Cervera and H. D. Ceniceros, Phys. Fluids **19**, 023101 (2007).

[164] J. Drappier, D. Bonn, J. Meunier, S. Lerouge, J.-P. Decruppe and F. Bertrand, J. Stat. Mech.: Theory and Exp. **4**, 04003 (2006).

[165] S. M. Fielding and P. D. Olmsted, Phys. Rev. Lett. **96**, 104502 (2006).

[166] S. M. Fielding, Phys. Rev. Lett. **95**, 134501 (2005).

[167] E. Fischer and P. T. Challaghan, Phys. Rev. E **64**, 011501 (2001).

[168] M. M. Britton, R. W. Mair, R. K. Lambert, and P. T. Callaghan, J. Rheol. **43(4)**, 897 (1999).

[169] D. Marenduzzo, E. Orlandini and J. M. Yeomans, J. Chem. Phys. **121**, 582 (2004).

[170] H. Zhou, M. G. Forest, and Qi Wang, Discrete and Continious Dynamical Systems-Series B **8 (3)**, 707 (2007).

[171] G. Kiss and R. Porter, J. Poly. Science: Poly. Symp. **65**, 193 (1978).

[172] S.-G. Baek, J. J. Magda, and D. Cementwala, J. Rheol. **37**, 935 (1993).

[173] S. Heidenreich, S. Hess, S. H. L. Klapp, M. Gregory Forest, Ruhai Zhou, and Xiaofeng Yang, AIP Conf. Proc. **1027**, 168 (2008).

[174] S. Heidenreich, S. Hess, and S. H. L. Klapp, Phys. Rev. Lett. **102**, 028301 (2009).

[175] J. Harden, B. Mbanga, N. Eber, K. Fodor-Csorba, S. Sprunt, J. T. Gleeson, and A. Jakli, Phys. Rev. Lett. **97**, 157802 (2006).

[176] L. Bocquet and J.-L. Barrat, J. Phys.: Condens. Matter **8**, 9297 (1996).

[177] D. Grecov and A. R. Rey, Phys. Rev. E **68**, 061704 (2003).

[178] J. L. McWhirter and G. N. Patey, J. Chem. Phys. **123**, 084502 (2005).

[179] H. Zhou, H. Wang, Q. Wang and M. G. Forest, Nonlinearity **20**, 277 (2007).

[180] G. Barbero and Y-.A Kosevich, Phys. Lett. A **170**, 41 (1992).

[181] A. V. Zakharov and R. Y. Dong, J. Chem. Phys. **116**, 6348 (2002).

[182] H. R. Brand, H. Pleiner, and F. Ziebert, Phys. Rev. E **74**, 021713 (2006).

[183] H. Pleiner and H. R. Brand, *In Pattern formation in liquid crystals: Hydrodynamics and Electrohydrodynamics of liquid crystals*, Springer N.Y., 1996.

BIBLIOGRAPHY

[184] E. Jarkova, H. Pleiner, and H.-W. Müller, J. Chem. Phys. **118**, 2422 (2003).

[185] H. B. Brand and H. Pleiner, Physica **208A**, 359 (1994).

[186] G. Barbero, G. Skacej, A. L. Alexe-Ionescu and Zumer, Phys. Rev. E **60**, 628 (1999).

[187] J. Jordanovic and S. H. L. Klapp, Phys. Rev. Lett. **101**, 038302 (2008).

[188] G. M. Whitesides, Nature **442**, 368 (2006).

[189] J. Yang, F. Lu, L. W. Kostiuk and D. Y. Kwok, J. Micomech. Microeng. **13**, 963 (2003).

[190] S. Hess, Zeitschrift f. Naturforschung **35a**, 69 (1980).

[191] I. Pradowitz and S. Hess, Physica **100A**, 540 (1979).

[192] S. Hess, Physica **314A**, 310 (2002).

[193] L. Manna, D. J. Milliron, A. Meisel, E. C. Scher, A. P. Alivisatos, Nature Materials **2**, 382 (2003).

[194] R. Voituriez, J. F. Joanny and J. Prost, Phys. Rev. Lett. **96**, 028102 (2006).

[195] S. Ramaswamy and M. Rao, New. J. Phys. **9**, 423 (2007).

[196] S. Ramaswamy and R. A. Simha, Solid State Commun. **139**, 617 (2006).

[197] Y. Hatwalne, S. Ramaswamy, M. Rao and R. A. Simha, Phys. Rev. Lett. **92**, 1181101 (2004).

[198] S. Muhuri, M. Rao, and S. Ramaswamy, Eur. Phys. Lett. **78**, 48002 (2007).

[199] M. E. Cates, S. M. Fielding, D. Marenduzzo, E. Orlandini, and J. M. Yeomans4, Phys. Rev. Lett. **101**, 068102 (2008).

[200] H. Chaté, F. Ginelli and R. Montagne, Phys. Rev. Lett. **96**, 180602 (2006).

[201] T. B. Liverpool and M. C. Marchetti, *Hydrodynamics and rheology of active polar filaments*, Biological and Medical Physics, Biomedical Engineering, Springer New York, 2008.

[202] T. B. Liverpool and M. C. Marchetti, Phys. Rev. Lett. **97**, 268101 (2006).

[203] A. Baskaran and M. C. Marchetti, Phys. Rev. E **77**, 011920 (2008).

[204] A. Ahmadi, T. B. Liverpool and M. C. Marchetti, Phys. Rev. E **72**, 060901 (R) (2005).

[205] D. Marenduzzo, E. Orlandini, M. E. Cates, and Yeomans, Phys. Rev. E **76**, 031921 (2007).

[206] D. Marenduzzo, E. Orlandini and J. M. Yeomans, Phys. Rev. Lett. **98**, 118102 (2007).

BIBLIOGRAPHY

[207] D. Marenduzzo, E. Orlandini, and J. M. Yeomans, Phys. Rev. Lett. **98**, 118102 (2007).

[208] V. Narayan, S. Ramaswamy, N. Menon, Science **317**, 105 (2007).

[209] I. Llopis and I. Pagonabarraga, Europhys. Lett. **75**, 999 (2006).

[210] L. Lapidus and G. F. Pinder, *Numerical Solutions of Partial Differential Equations in Science and Engineering*, Wiley-Intersience Publication, Toronto, 1982.

[211] D. Marsal, *Die numerische Lösung partieller Differentialgleichungen in Wissenschaft und Technik*, Bibliographische Institut Zürich, 1976.

[212] R. E. Mickens, *Nonstandard finite difference models of differential equations*, World Scientific, 1994.

VDM Verlagsservicegesellschaft mbH

Die VDM Verlagsservicegesellschaft sucht für wissenschaftliche Verlage abgeschlossene und herausragende

Dissertationen, Habilitationen, Diplomarbeiten, Master Theses, Magisterarbeiten usw.

für die kostenlose Publikation als Fachbuch.

Sie verfügen über eine Arbeit, die hohen inhaltlichen und formalen Ansprüchen genügt, und haben Interesse an einer honorarvergüteten Publikation?

Dann senden Sie bitte erste Informationen über sich und Ihre Arbeit per Email an *info@vdm-vsg.de*.

Sie erhalten kurzfristig unser Feedback!

VDM Verlagsservicegesellschaft mbH
Dudweiler Landstr. 99
D - 66123 Saarbrücken

Telefon +49 681 3720 174
Fax   +49 681 3720 1749

www.vdm-vsg.de

Die VDM Verlagsservicegesellschaft mbH vertritt

Printed by Books on Demand GmbH, Norderstedt / Germany